DATE DUE

IN THE LIGHT OF THE SUN

FROM SUNSPOTS
TO SOLAR ENERGY

IN THE LIGHT
OF THE SUN

MARK WASHBURN

Harcourt Brace Jovanovich
New York and London

Requests for permission to make copies of any
part of the work should be mailed to:
Permissions, Harcourt Brace Jovanovich, Inc.,
757 Third Avenue, New York, N.Y. 10017

Library of Congress Cataloging in Publication Data

Washburn, Mark.
In the light of the sun.

Bibliography: p.
Includes index.
1. Sun—Popular works.
2. Solar energy—Popular works.
I. Title.
QB521.4.W37   523.7   80-8762
ISBN 0-15-186737-2

Printed in the United States of America

First edition

B C D E

For my brothers,
Rusty and Scott,
and for my nephews,
George and Geoff

Contents

CONTENTS

Acknowledgments

This book would have been much easier to write if all the solar scientists had agreed to a moratorium on further research. They didn't, alas, but many of them did take time out to discuss their work with me. Most of those conversations occurred during the summer and fall of 1979; between then and now, their work has continued, and it is likely that at least some of the information attributed to specific persons may have been superseded by more recent findings. It is possible that a scientist who advocated Theory A when I spoke with him may now support Theory B— things change rapidly in the solar sciences. Nevertheless, I believe that the information presented here accurately reflects the state of the art at the end of the 1970s. I am grateful to all the many persons who contributed to this book; if any misinformation is presented here, it is my fault, not theirs.

Specifically, my sincere thanks go to the following: at the Solar Energy Research Institute (SERI), Bruce Baccei, Larry Douglas, Dan Jantzen, Ralph Kerns, Chuck Kutscher, and Arnon Levary; at the National Center for Atmospheric Research (NCAR), Gordon Newkirk, Raymond Roble, Stephen Schneider, and Dick White; at the National Oceanic and Atmospheric Administration (NOAA), Gary Heckman and George Reid; and at the Aspen Institute, Roger Olson. For arranging all these interviews, I am extremely grateful to Jerome Williams (SERI), Dianne Johnson and Ralph Segman (NCAR), Bonnie Rodriguez (Kitt Peak), and Carl Posey (NOAA). Thanks also to John Bahcall (Princeton), A.G.W. Cameron (Harvard), Nevin Bryant (JPL), and Keith Pierce (Kitt Peak).

Special thanks to John A. Eddy (NCAR) for his many valuable contributions to this book.

My fellow journalists also merit thanks for their unselfish and enthusiastic literary, scientific, and logistical support. I am especially grateful to David Salisbury, J. Kelly Beatty, and Jonathan Eberhart.

My peerless agents, Helen Brann and John Hartnett, made this book possible in the first place. My equally peerless editors, Brian Dumaine and Carol Meyer, have restored my faith in the entire field of editing and publishing.

Finally, there are a great many people who deserve thanks for their support and encouragement: Bob Webb, Mike Germer, Jon Lomberg, Brown, Shirley Arden (bionic executive assistant), Carl Sagan, Don Bane, Tom Campbell and Marci Kramish, Karen Nelson, Jack and Kathy Sheets, Tom Petruso, Cosmo Gia, the Shuck family, and my own family. And very special gratitude goes to neutrino expert Karen Keating and the many members of her clan. My thanks to them all.

Introduction

When I was in the fourth grade, my class went outside one day for an art exercise. Armed with crayons, we were instructed to draw anything we wanted to; I chose for my subject a nearby house which was still under construction. My artistic endeavor was proceeding smoothly until the construction workers returned from a coffee break. I watched in horror as they installed a new window and began painting the side of the house I had been drawing. My picture was becoming obsolete before my eyes. With a flush of preadolescent wisdom, I resolved never again to paint a portrait of a house under construction. Now, twenty-odd years later, I find that I've done it again.

I would like to be able to say that this book is a reasonably complete and up-to-the-minute picture of the state of our knowledge about the Sun. Unfortunately, it isn't. Today's solar science resembles nothing so much as an amoeba with a hyperthyroid condition; it wriggles and squirms, changes shape, divides, and multiplies, all without even a pause for a coffee break. It is a singularly uncooperative subject for a still-life portrait. Passages of this book written less than a month ago already need revision. The mechanics of book publishing guarantee that by the time this reaches print there will be need for further revisions.

This is not a new problem. In 1940, the Russian-American physicist George Gamow wrote a popular book about the Sun. In 1964 he wrote another book about the Sun, which was dedicated "To the memory of my book, *The Birth and Death of the Sun*," which by 1964 was almost completely obsolete. In a quarter of a century, solar science had undergone a sweeping revolution. Today, the rate of change is considerably more rapid.

The Sun is being examined from every conceivable vantage point. Data are being collected from the bottom of a gold mine and from a spacecraft beyond the orbit of Saturn, from computer models and medieval manuscripts, from oceanic sediments and Moon rocks. The mass of new knowledge about the Sun is overwhelming, but far from complete. Discoveries of the 1970s are already being overshadowed by the scientific harvest of the 1980s.

We are learning, not just about the Sun itself, but about its influence on the Earth and its inhabitants. Scientific disciplines which for centuries have been isolated enclaves are suddenly converging. A few years ago, a book such as this one would have been concerned almost exclusively with the work of astronomers and physicists. They are still well represented, but a partial list of other fields of study now impinging on solar science would include microbiology, meteorology, atmospheric chemistry, climatology, glaciology, geology, organic chemistry, high-energy physics, paleontology, oceanography, botany, comparative planetology, and even archaeology. Political and economic factors are also involved in the study of the Sun, and these, too, must be addressed.

In spite of all the attention now being paid the Sun, the blunt truth is that no one understands it. We are rather like college freshmen home on vacation, bubbling over with exciting new knowledge, but unsure how it all fits together. And like college freshmen, we are discovering unsettling chinks in supposedly eternal verities and baffling new questions about things we never knew existed.

No one can say where these discoveries will lead. Science used to march ahead at a stately pace; today it gallops. Some people find all these discoveries frightening. Others find them merely boring. But for those with the energy and courage to confront the universe in all its awesome complexity, this is an incredibly exciting time to be alive.

That excitement is the reason for this book. In these pages I have attempted to convey some of the sense of wonder and anticipation surrounding contemporary solar science. I have tried to be accurate and precise, but this is in no sense a definitive study of the Sun. Think of it, rather, as resembling a seventeenth-century map of the known world. The continents and oceans are more or less in the correct positions, and some of the rivers and mountain ranges have been plotted in detail, but there are still vast areas marked only: "Unexplored." Some of those blank spaces will be filled in before you read this; others will remain for your great-grandchildren to survey.

So here it is, for your entertainment and edification—a slightly blurred snapshot of a subject in rapid motion. May you enjoy reading it as much as I have enjoyed writing it.

IN THE LIGHT OF THE SUN

One: STARBIRTH

In the beginning there was the cloud.

Dark, silent, and cold, the cloud had existed for billions of years, floating undisturbed across the vast reaches of the universe. The cloud was immense, perhaps a hundred thousand billion miles in diameter, yet it was not unique. Other clouds, older and even larger, had already condensed to form the Milky Way galaxy, a colossal whirlpool of stars and dust that extended for a hundred thousand *trillion* miles.

Two-thirds of the way out from the center of the whirlpool, the cloud's domain was one of the graceful spiral arms that curved outward from the core of the young galaxy. Here, the infinitesimal debris of cosmic birth and death had collected, forming a dark lane between the glowing limbs of the galaxy.

The cloud was enormous, but it was thin. In each cubic centimeter, there were only a few thousand atoms. Almost all of the atoms were hydrogen, the simplest atom in nature: a single, wispy electron whirling endlessly about a naked proton. Yet each atom exerted a minute tug on all the others in the cloud, and in concert they performed a delicate gravitational dance that gave the cloud form and structure.

At the center of the cloud, the concentration of atoms was greater, the pull of gravity stronger. More and more atoms gathered there. But as the attracting power of the core increased, it was balanced by another force. The electrons of the hydrogen atoms occasionally shifted their orbits around the protons, and in the process released a tiny burst of energy. It was a rare event, occurring only once in about eleven million years for each hydrogen atom. At the core of the cloud, however, there were so many atoms that energy was released continuously.

These hydrogen emissions gradually raised the temperature inside the cloud. By human standards it was still forbiddingly cold—only ten degrees above absolute zero—but it was nevertheless warmer than the outer regions of the cloud. The thermal pressure flowing outward balanced the gravitational pull inward. The cloud became stable.

The cloud may have endured in this equilibrium for billions of years. Had it been alone in space, its stability might never have been disturbed.

But four and a half billion years ago something happened. Somewhere in the same spiral arm as the cloud, a star died in a sudden apocalyptic fury, releasing prodigious amounts of energy in the course of hours, rather than eons. The remnants of the exploding star swept past the cloud, infusing it with some of the enormous energy of the supernova.

This external energy overwhelmed the pressure emanating from the center of the cloud, and the balance of pressure and gravity was broken. The cloud began to collapse inward upon itself.

The cloud's collapse set in motion a chain of events that ultimately led to three extremely important results: the Sun, the Earth, and Man. Our heritage is identical with that of the planet we stand upon and the star that warms us. The raw materials of each were contained in that swirling, tenuous cloud, four and a half billion years ago. One might say that it has taken that long for the cloud to become aware of itself.

Long before Man had any notions about hydrogen clouds and solar nebulae, he was well aware of his connection with the Sun. The link was obvious, even to the primitive hunter-gatherers who created civilization some fifteen to twenty thousand years ago. The Sun was the source of heat and light, without which life would be impossible. Men began to study the motions and behavior of the Sun and, perhaps inevitably, to worship it.

Nearly every ancient culture engaged in Sun worship to some degree. The Sun god Utu was of supreme importance to the Sumerians; for the Babylonians, Utu became Marduk; for the Egyptians, Ra. The Aztecs worshiped a particularly bloodthirsty Sun god, and the Incas thought of themselves as the Children of the Sun. Even in modern religions we can hear the echoes of the ancient Sun worshipers. The concept of monotheism certainly

In this ancient view of the sun, the earth is surrounded by concentric spheres inhabited by the stars, sun and moon, water, and fire.
New York Public Library Picture Collection

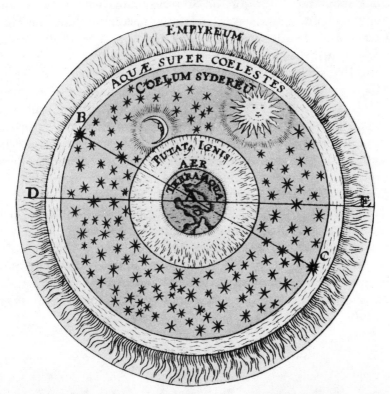

derives in part from Sun worship—there was only one Sun, and thus only one God. The religious symbol of the cross is also, to an extent, a holdover from the early Sun religions; the use of the cross as a religious symbol predates Christianity and was originally intended as a representation of the rays of the Sun.

A considerable portion of the social energy of early civilizations was devoted to building structures to honor and keep track of the Sun. The megalithic structures at Stonehenge in England and the Big Horn Medicine Wheel in Wyoming are now known to be complex astronomical computers designed to chart the motions of the heavens. The Pyramids of Egypt served a similar function, although they were not quite so miraculously precise as some writers would like to believe. The ancients, of course, had no idea what the Sun really was or why it behaved as it did. They recorded its visible manifestations, but when the Sun went down, imagination took over. The Egyptians believed that after the Sun set in the west, it returned to the east by way of a demon-infested underground cavern.

HOW BIG, HOW FAR?

The ancient Greeks were probably the first to ask serious questions about the nature of the Sun, beginning with how big the Sun is and how far away it is. Definitive answers to these questions were not achieved until the twentieth century, but the Greeks came up with some surprisingly good estimates.

In the sixth century B.C., the astronomer Anaximander of Miletus wrote that the Sun "is like a chariot wheel, the rim of which is hollow and full of fire, letting the fire shine out at a certain point in it like the nozzle of a pair of bellows." In the same era, Heraclitus of Ephesus declared that the Sun could be no more than about twelve inches in diameter. Ludicrous as these early theories may sound, they represented attempts to describe the Sun in quantifiable, nonmythical terms. At the very least, they provided a point of departure for better theories. One such theory was put forward a century later by Anaxagoras, who suggested that the Sun was really a fiery rock about thirty-five miles in diameter. At about the same time, Philolaus of Tarentum became the first to suggest that the Earth was really a sphere floating in space. The idea soon took hold, and was never again seriously challenged.

Given two independent spheres, size and distance could be calculated through simple geometry. Aristarchus of Samos in the third century B.C. performed the necessary exercises and concluded that the Sun was some 720,000 miles away from the Earth. If it was that far away, the Sun had to be at least as large as the Earth. In the next century Hipparchus of Nicaea, known as the father of modern astronomy, recalculated the problem and came up with a distance to the Sun of about five million miles. The Sun therefore was considered to be about seven times the size of the Earth.

Hipparchus's answers to the fundamental questions were wrong, but the achievements of the ancient Greeks should not be minimized. In everyday life, one never encounters five million of anything; certainly, in ancient Greece, it required some intellectual courage to imagine that anything—even the Sun—could be five million miles away. The size of the universe and the scope of Man's thinking had been broadened significantly.

Unfortunately, the physical and intellectual expansion of the universe came to a halt following Hipparchus. For eighteen centuries, the prevailing wisdom held that the Earth was the center of creation, and the Sun and everything else revolved around it. Aristarchus had suggested that perhaps it was the Sun that was in the center, but his heliocentric theory was submerged under the considerable weight of Aristotelian orthodoxy. The Earth (and science) stood still until the sixteenth century A.D. when the Polish astronomer Nicolaus Copernicus proposed his own heliocentric theory.

The Copernican Revolution opened the floodgates of scientific inquiry. Sixty years after the publication of Copernicus's theory, the German Johannes Kepler worked out his laws of planetary motion. Kepler's careful study of the motions of the planet Mars showed that planetary orbits around the Sun are elliptical and that a planet's distance from the Sun is related to the time it takes to complete its orbit. Knowing this, it was possible to calculate the relative distances of each of the planets. Thus, if the true distance of any one of the planets could be found, the dimensions of the entire solar system could be calculated.

In order to find the distance to a planet, it was necessary to observe a parallax—the apparent shift in position of a distant object when viewed from different locations. The first good measurement of a planetary parallax was made in 1671 by Jean Richer and Giovanni Domenico Cassini. Richer led an expedition to French Guiana, while Cassini remained in Paris. Each made careful observations of Mars relative to the presumably immobile background stars. Comparing the results, Cassini was able to calculate the distance to Mars and, by Kepler's laws, the distance to the Sun. His result: 87,000,000 miles.

A century later, Cassini's figures were improved by another scientific expedition. The British Navy dispatched an obscure lieutenant named James Cook to the South Pacific to observe the 1769 transit of Venus across the face of the Sun, in the hope of obtaining a very precise parallax for that planet. Cook succeeded in observing the transit; on his way home, he also observed New Zealand and Australia, an example of the sometimes serendipitous results of scientific research.

The data gathered by Cook and others later enabled German astronomer Johann Franz Encke to calculate a distance to the Sun of 95,370,000 miles. Encke's result is interesting because it was probably the first time the distance to a celestial object had been *over*estimated.

It was not until the twentieth century that astronomers finally zeroed in on the actual distance to the Sun. In 1931, meticulous observations of the orbit of the asteroid Eros led to the determination of an average distance to the Sun of slightly less than 93,000,000 miles. Since the Earth, like the other planets, has an elliptical orbit, the distance to the Sun varies between about 91,400,000 miles (perihelion) and 94,600,000 miles (aphelion). The parallax method has recently been improved upon by modern technology. In 1961, scientists bounced microwave radar beams off Venus. By measuring the time it took for the beams to return to Earth, an exact measurement of distance was possible. The Venus results translated into an average Earth-Sun distance of 92,960,000 miles. The Sun was found to have a diameter of approximately 864,000 miles, making it 109 times the size of the Earth. After some three thousand years, the questions of how big and how far had finally been answered.

As astronomers found answers, they also discovered more questions. From ancient times until the modern era, it was assumed that the stars were fixed points of light, possibly attached to some grand celestial shell which contained the universe. But some stars are brighter than others, and it occurred to astronomers that the differences in brightness could be the result of varying distances. This led to the intriguing notion that the stars were simply other suns, seen from very far away. Or, to put it another way, the Sun was simply a star seen from nearby. Aristarchus may have been the first to suggest that this was the case.

There were problems, however, with the theory that the stars were distant suns. The stars were apparently unchanging and immobile, unlike everything else in the heavens. They also seemed to be very small, since even the telescope showed them to be no more than pinpoints of light. Finally, the stars showed no parallax. The great Danish astronomer Tycho Brahe rejected the Copernican theory on the basis of this point alone, believing that if the Earth were in motion, the stars would have to show at least some parallax. An occasional heretic, such as the Italian philosopher Giordano Bruno, might speculate that the stars were distant suns, surrounded by unseen planets and civilizations, but science could find no evidence to support such a theory.

In the eighteenth century, the knot finally began to unravel. British astronomer Edmund Halley studied the star maps plotted by Hipparchus in 134 B.C. and found that three prominent stars were no longer in the positions recorded by the ancient Greek. Observational error seemed to be ruled out because both Hipparchus and Halley were meticulous observers. Halley found that the majority of stars checked out perfectly with the positions plotted on Hipparchus's ancient star chart, but that Arcturus, Procyon, and Sirius had each moved by as much as twice the apparent width of the full moon. Evidently the stars were not fixed, but possessed their own "proper motions" through the heavens. The idea was supported by the fact that the three stars in question are among the brightest in the sky. If they were bright, they might be relatively nearby, which would make their motions more obvious than those of more distant stars.

The fact that the stars moved did not, in itself, prove that the stars were suns and the Sun a star. The stars might, for example, be very small, slow-moving objects just beyond the visible planets. Proof would come only when the distance to the stars could be established, which meant that astronomers needed to detect a stellar parallax.

The search for a stellar parallax continued for centuries. It was a frustrating quest because of the great precision necessary to make such a measurement. But the need for precision spurred the evolution of better astronomical equipment and techniques. Finally, in the late 1830s, three different astronomers discovered indisputable stellar parallaxes. Their work led to the calculation that Alpha Centauri, which turned out to be the nearest star, was some twenty-five trillion miles away from the Earth. The immensity of this number was immediate confirmation that the stars were, indeed, suns in their own right. Our own Sun, when viewed from such a distance, wouldn't even be the brightest star in the sky.

Astronomers find it convenient to have a star nearby. Aside from the obvious benefits, the presence of the Sun enables astronomers to make a relatively close-up examination of a universal phenomenon. We cannot see what is happening on the surface of Sirius or Rigel, but we can quite easily observe events on the surface of the Sun. When a solar flare erupts or the number of sunspots increases, we know about it almost immediately. We are unlikely to observe similar events on Sirius no matter how carefully we watch it, but we can nevertheless safely assume that they do happen.

But is that really such a safe assumption? Can we be certain that what we see happening on the Sun is typical of stars in general? Is it valid to extrapolate data from one star to billions?

One way of approaching the problem is to look not at the tree, but the forest. In this case, the "forest" consists of hundreds of billions of stars in our own galaxy and uncountable trillions more in a hundred billion other galaxies. If our own Sun is thought of as a sturdy nearby oak tree, then our study of it can give us only a limited amount of information about trees of a particular species, age, and size. But out in the forest, we can observe not only oaks, but birch and elm and redwoods, saplings, fallen logs, and even an occasional forest fire. We can, perhaps, understand our own oak tree in terms of what we can learn about trees in general.

A HUNDRED BILLION SUNS

Out in the galactic forest, we can see an astonishing number of trees. Estimates of the number of stars in the Milky Way galaxy vary; the lower limit is about 100 billion and the upper is about 250 billion. We have difficulty in obtaining better accuracy because toward the center of the galaxy the stars are so numerous that they blot out our view of what lies beyond.

Even without the aid of a telescope, it is clear to the casual observer that there are differences among the stars. The most obvious differences are in brightness and color. On any given night, there are likely to be about a half-dozen stars that are much brighter and more prominent than the rest. Significantly, there are many more stars which are not so bright, and as brightness decreases, the population increases. This was a clue, even to ancient observers, that there might be large differences in the distances to the stars.

Color is a more subjective matter, but there is no doubt that differences do exist. The majority of stars seem to range from yellow to white to blue. Among the brightest stars, Sirius, Vega, and Rigel appear bluish-white. Capella has a slight yellowish cast to it, not unlike the Sun. Some bright stars, such as Aldebaran, Arcturus, and Betelgeuse, are distinctly red or orange.

These obvious differences in brightness and color served as a starting point for the study of the stars. Hipparchus was the first to classify stars according to brightness. He assigned each star to one of six groups, or magnitudes; the brightest stars were of first magnitude and the dimmest ones visible were of sixth magnitude.

Modern astronomers have retained Hipparchus's scheme of magnitudes, but have refined it to a mathematical system. A given magnitude is 2.512 times as bright as the next lower magnitude; thus, a first magnitude star is 100 times as bright as a sixth-magnitude star. The brightest stars require zero or negative magnitudes. Sirius, the brightest star in our sky, has a magnitude of -1.4. The system also applies to nonstellar objects. The planet Venus can be as bright as -4.3, while the full moon registers at -12.6. The Sun has a magnitude of -26.9. In the other direction, telescopes have revealed stars and other objects with magnitudes far beyond the naked-eye limit of 6. The most distant galaxies have magnitudes of about 23.5.

Once it was established that there are great differences in the distances to the stars, it was clear that the system of magnitudes would have to be revised. We were recording, after all, only the apparent magnitudes of the stars as seen from Earth. In order to determine the actual intrinsic brightness of a star, another system is needed.

To establish "absolute magnitudes," astronomers arbitrarily "move" stars to a standard distance of 10 parsecs, or 32.6 light-years, away from the Earth.* By this standard, Sirius, which is only 8.7 light-years away, has an apparent magnitude of −1.4, but an absolute magnitude of only +1.4. Canopus, on the other hand, has an apparent magnitude of −0.7, but an absolute magnitude of −3.1. The Sun, resplendent in the sky at −26.9, turns out to have an absolute magnitude of only +4.77; if the Sun were 32.6 light-years away, it would barely be visible to the naked eye.

The system of absolute magnitudes made it clear that the apparent differences among the stars were not simply the effects of varying distances. The differences were intrinsic and real. But why should that be the case? And if the Sun was a star, what could account for the differences between it and the other stars?

STARLIGHT

If we were left to comparisons of intrinsic brightness, we would have a major mystery. Fortunately, the universe is structured in such a way that a single beam of light contains an astonishing amount of information. Brightness is only one of the messages conveyed by starlight; there are many others.

Sir Isaac Newton founded the science of optics with his discovery, reported to the Royal Society in 1672, that sunlight, when directed through a prism, breaks down into a spectrum of colors ranging from red to violet. Two centuries later, Joseph von Fraunhofer, Robert Wilhelm Bunsen, Gustav Robert Kirchoff, and other scientists examined the spectrum more closely and found a wealth of detail in Newton's rainbow. There is much more to light than meets the eye.

Light can be thought of as an undulating wave. The visible color of a beam of light depends on the length of the waves; red light consists of longer waves than violet. Our eyes are sensitive only to the limited range between red and violet, but both longer

* The sky is divisible into 360 degrees of arc, with 60 minutes to each degree and 60 seconds per minute. The full moon, for example, has an apparent diameter of 31 minutes of arc; 1 second of arc is thus equal to about 0.0005 of the lunar diameter. In order for a star to show a parallax of 1 second of arc, it would have to be 3.26 light-years away. Thus, 1 parallax second, or "parsec," is equal to a distance of 3.26 light-years. A light-year is the distance traveled by a beam of light in one year— roughly 6 trillion miles.

and shorter wavelengths exist. In 1861, the great Scottish physicist James Clerk Maxwell proposed that all light is simply a part of a larger electromagnetic spectrum. At one end of the spectrum, there are extremely long and not very energetic waves which can be detected by your car radio and your television set. Slightly shorter, more energetic waves cook your food in microwave ovens. Beyond microwaves, the broad infrared region begins, followed by visible light and then ultraviolet. Still shorter, extremely energetic waves can be detected in the form of x-rays and gamma rays.

Nineteenth-century scientists, for a variety of reasons, were more or less limited to studying the visible wavelengths of light, but it is unlikely that they felt deprived. Early in the century, Fraunhofer found that Newton's rainbow contains numerous dark lines. By midcentury, Kirchoff and Bunsen discovered that Fraunhofer's dark lines corresponded to bright lines in the spectra produced by incandescent substances in the laboratory. It seemed that each element possessed its own characteristic spectral lines, and the appearance of those lines in the solar spectrum implied the presence of those elements in the Sun.

The Sun produces a continuous spectrum of visible light. As that light, in the form of photon particles, makes its way outward from the core, it interacts with the various elements present in the Sun. Some elements absorb energy at specific wavelengths, subtracting bits and pieces from the total spectrum, resulting in the dark Fraunhofer, or absorption, lines. Other elements, when heated, may release energy at given wavelengths, producing in the spectrum bright lines known as emission lines.

These discoveries opened up an entirely new field of study: spectroscopy. One could study the spectra of the Sun and stars and deduce their composition. In 1862, Swedish astronomer A. J. Ångstrom discovered the characteristic lines of hydrogen in the solar spectrum. Kirchoff had already found those of sodium and calcium. Similar discoveries were made from the spectra of the stars. Apparently the visible universe was made of the same stuff as the Earth itself. But not all of the lines in the solar spectrum were readily identifiable. British astronomer J. Norman Lockyer theorized that certain unidentified lines were the signature of a previously unknown element, which he called helium, after the Greek Sun god Helios. Not until several years later was helium detected on Earth. The technique, however, was not infallible.

For years, astronomers believed in the existence of two other non-earthly elements—"coronium" in the Sun's outer atmosphere, and "nebulium" in distant clouds of gas. These supposed "discoveries" were seriously to mislead solar astronomers.

By the beginning of the twentieth century, astronomers had analyzed the spectra of thousands of stars. Inevitably, patterns emerged from the data. At Harvard, E. C. Pickering and a small group of assistants attempted to create a logical system of spectral classification, in which the prominent features of one spectral group would blend into the characteristics of the next group. Since hydrogen seemed to be present in every star, they based their system on the intensity of the hydrogen lines in stellar spectra, starting with the strongest and ending with the weakest.

The Pickering group found, however, that basing the system on hydrogen caused the spectral lines of other elements to become jumbled. The groups were eventually rearranged to make the most sense of the data, resulting in the present unalphabetical sequence of spectral classes: O, B, A, F, G, K, M, R, N, S. Generations of astronomy students have learned the sequence from the mnemonic: "Oh, Be A Fine Girl, Kiss Me Right Now, Sweet." For fine tuning, each class was subdivided with numbers ranging from 0 to 9. In this scheme, the Sun is classified as a G0 star.

Significantly, the spectral classes corresponded with the visible colors of the stars. Bright blue stars, such as Rigel, are found in class B. White stars, such as Sirius, are in class A. Capella, a yellowish star, is, like the Sun, a G0. The orange star Arcturus is a class K, and red stars, such as Antares, are grouped in class M.

The spectrum of a class O star is the simplest, and is dominated by strong hydrogen lines, which peak at class A0 and then fade. Helium is also prominent in O and B stars. Beginning in class A, the lines of metallic elements such as calcium and iron become important, reaching maximum strength and complexity in F, G, and K stars. M, R, N, and S stars have extremely complicated spectra which are dominated by the signatures of various molecular compounds.

Superficially at least, classifying stars by their spectra would seem to be pretty straightforward. In reality, a variety of complications come into play. Spectral lines are seldom sharp and distinct; more often, they are broad, fuzzy, and displaced, making them much more difficult to "read." In the well-known Doppler effect, discovered by Austrian physicist Christian Johann Doppler in 1842, spectral lines shift according to the motion of the star (and of the observer); objects moving toward the observer are shifted toward violet, while receding objects display "red shifts." In addition, there are the Zeeman effect—the splitting of spectral lines due to interactions with magnetic fields—and the Stark effect —splitting due to electrical fields. Further complications are introduced by hyperfine structures within the spectral lines, turbulence and expansion in stellar atmospheres, and the rotation of stars.

Complications aside, the correspondence between spectral classes and color is obvious. Two possible explanations for the spectral sequence occurred to astronomers. Since different elements showed up in the spectra of different stellar classes, it was possible that the color differences were the result of chemical composition. Different chemicals produce flames of differing colors, a fact known for centuries to makers of fireworks. Might a class M star appear red because it was made of different elements than a class O star?

It might—but a better explanation is that a class M star is burning at a lower temperature than a class O star. A blacksmith heating a horseshoe will see the iron begin to glow with a dull red color; as it becomes hotter, it turns orange, then yellow, and finally, white hot. Stars do essentially the same thing.

Measuring the temperature of a star is a difficult procedure because of a host of complications having to do with distance, size, structure, interference by the Earth's atmosphere, and details of quantum physics. Nevertheless, stellar temperatures can be calculated if the distance and size of a particular star are known. Distance can be determined by the parallax method (and other more sophisticated techniques), and the size of some stars can be calculated by the use of a device known as the stellar interferometer, invented by American physicist Albert A. Michelson. Brightness is an observed quantity. Taken together these measurements can be used to calculate some relatively accurate values for the surface temperatures of some stars. The data can then be extrapolated to other similar stars.

By the twentieth century, it was becoming clear that the galactic forest was a place of incredible diversity. Stars were small or huge, hot or cool, stable or volatile. But what could account for such a sweeping variety of stars? If all stars were composed of the same inventory of raw materials, shouldn't they all be pretty much alike? Perhaps the answer lay in the method by which stars form in the first place.

Stars, of course, are not the only denizens of the cosmos. As telescopes improved, astronomers began to take note of hundreds of hazy, luminous patches which they called nebulae. These seemed to consist of vast regions of unconsolidated dust and gas. They glowed from the light of nearby stars, yet they were clearly not stars themselves. Nebulae were apparently quite common in the universe; perhaps they played a role in the formation of stars and planets.

THE SOLAR NEBULA

In the eighteenth century, German philosopher Immanuel Kant and French astronomer Pierre Simon de Laplace both thought about the problem and independently arrived at similar theories. They suggested that a large cloud of dust and gas would eventually contract inward upon itself, in accord with the laws of gravitation. As a nebula contracted, it would spin faster and faster until centrifugal forces finally caused it to shed successive layers of matter. These layers would form flat rings around the center of the nebula. In time, the core of the cloud would become a star, and the ejected matter would condense into planets.

This theory was attractive for a number of reasons. It explained why the planets were all more or less in the same plane and rotated in the same direction.* It also implied that if all stars formed in the same manner, then planets—and probably life—were very likely ubiquitous.

Throughout the nineteenth century, the Kant-Laplace nebular hypothesis was widely accepted. Still unanswered, however, was another disarmingly simple, childlike question: what makes the Sun shine? More specifically, what is the source of energy which permits the Sun and other stars to release such enormous quantities of heat and light?

Since ancient times, men had made the reasonable assumption that the Sun was burning, but by the nineteenth century scientists were asking *what* it burned. Even before spectroscopy permitted

* Actually, Venus and Uranus have a retrograde rotation.

a determination of the composition of the Sun, physicists such as Stefan, Boltzmann, Wein, and others were able to make rough calculations of the amount of energy being released at the Sun's surface: about 1,500 calories per square centimeter per second. German physicist Hermann Ludwig Ferdinand von Helmholtz compared that with the amount of energy liberated by the burning of coal and oxygen. He calculated that if the Sun were made of coal, it could be no more than about 2,000 years old; if it were older, it would be a burned-out cinder by now.

Clearly, the Sun is more than 2,000 years old, so it could not be a coal burner. Helmholtz began looking for alternative energy sources.

In 1853, Helmholtz came up with what looked like a good answer to the problem. If the Sun and its planets had truly formed from the contraction of the Kant-Laplace nebula, then perhaps the Sun released its gravitational energy in the form of heat. Working backward from the present size of the Sun to the presumed size of the original cloud, Helmholtz calculated that the Sun could have released a total of about 10^{41} calories during its lifetime. Assuming a constant rate of contraction and energy loss, Helmholtz concluded that the nebula began its contraction about 30 million years ago. The cloud would have shed the ring of matter destined to become the Earth some 18 million years ago. The Sun's continued contraction during recorded history would have amounted to a loss of only 560 miles from a diameter of 864,000 miles—not enough for anyone to notice.

In one brilliant exercise, Helmholtz seemed to have answered fundamental questions about the age of the solar system, the age of the Earth, and the nature of the Sun's energy supply. His was a very tidy theory, and it appealed to astronomers and physicists. The only problem was that geologists and biologists were already insisting that the Earth was much more than 18 million years old.

The Scottish geologists James Hutton and Charles Lyell had already laid the foundations of their discipline with their theory of uniformitarianism. They said that the geologic processes at work today—volcanism, erosion, sedimentation, and so forth—had been working continuously, at about the same rate, ever since the beginning. And in order to account for mountain ranges, oceans, and layers of sediment, the beginning must have come much earlier than 18 million years ago.

Soon after Helmholtz announced his theory, Charles Darwin came along and complicated the situation even further. Darwin's theory of the evolution of species declared that organisms evolve slowly, gradually adapting to make a better living from their environment. Again, 18 million years simply wasn't enough time to account for the multiplicity of organisms presently inhabiting the Earth. The biologists and geologists needed more time than the physicists and astronomers were willing to give them.

The controversy continued, unresolved, into the twentieth century. Some scientists thought that, if biological evolution had confused the question, perhaps a better understanding of stellar evolution might settle it.

In 1913, the study of the stars was aided by the introduction of a new analytical tool, the Hertzsprung-Russell diagram. The Danish astronomer Ejnar Hertzsprung had examined stellar temperatures and luminosities and concluded that while the hottest stars were always among the brightest, cool stars could be either bright or dim. To Hertzsprung, this indicated extreme variations in the size of the stars. Building on Hertzsprung's work, American astronomer Henry Norris Russell at Princeton devised what came to be called the Hertzsprung-Russell diagram.

The H-R diagram was simply a plot of absolute magnitude versus spectral class. Even Russell himself wasn't sure of its exact meaning. Students described it as "the plot of H against R."

When luminosity is plotted on the vertical scale, versus spectral class on the horizontal scale, the majority of stars fall into a narrow, slightly swaybacked band running from upper left to lower right, which is known as the Main Sequence. Hot, blue (O and B) stars are at the upper left of the band, and cooler red stars (K and M) are at the lower right. The Sun is just about in the middle of the band.

But not all stars are confined to the Main Sequence. Below it, to the left, there are a number of small but highly luminous stars known collectively as white dwarfs. Above the Main Sequence, to the right, are the giants and supergiants, red, cool, and immense.

It was obvious that the H-R diagram was showing something fundamental about stars, but in the beginning no one was quite sure how to decipher the message. At the time, Helmholtz's gravitational contraction theory of stellar formation was still in vogue. Working from it, Russell suggested that his diagram portrayed the life cycle of a star.

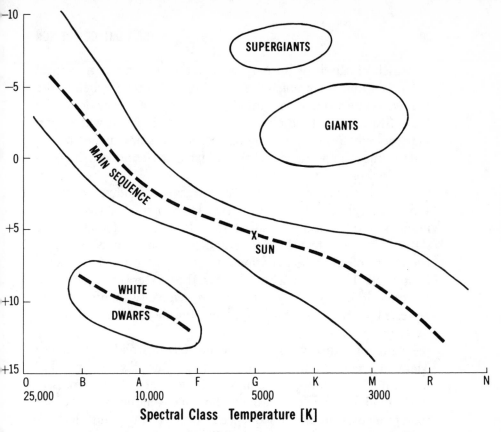

Spectral Class   Temperature [K]

Named after its creators, Ejnar Hertzsprung and Henry Norris Russell, the H-R Diagram is a highly useful tool for analyzing stars. Nearly all stars can be found in the broad band known as the Main Sequence. The Sun, spectral type G0, with a surface temperature of about 6000°K and an absolute magnitude of +4.77, can be found just about in the middle of the Main Sequence. It is a very average star. Eventually, in perhaps eight to ten billion years, it will expand dramatically and move toward the upper right of the H-R diagram, domain of the cool red giants and supergiants.

A star would begin life as a vast, barely luminous cloud of gas to the right of the Main Sequence. As it contracted, it would slowly move to the left along the top of the graph. Here, it would be in the red giant or supergiant stage. Eventually, the star would reach the top of the Main Sequence, as continued contraction caused it to burn white hot. At this point, the star enters the Main Sequence; it continues to shrink, but instead of getting hotter, it gets cooler, perhaps because the contraction process has some upper limit as far as luminosity is concerned. From here on, the star simply slides down the Main Sequence until it reaches the lower right as a dim, shrunken prune of a star, cold and dark.

Russell's analysis of his diagram had a certain aesthetic appeal but posed obvious problems. According to Russell's scheme, the Sun would be a star in late middle age, already halfway down the Main Sequence slide to oblivion. But by the 1920s, it was known that the Sun consisted mainly of an enormous amount of hydrogen. If the Sun were really middle-aged, it should not have so much hydrogen left.

In 1924, British astronomer Arthur Stanley Eddington found a possible solution to the problem. He proposed a basic mass-luminosity relationship in all stars. Eddington said that in any stable star, the internal pressure and temperature must be great enough to offset the gravitational forces pushing the star inward upon itself. The more massive the star, the greater the gravity, and therefore, the greater the temperature necessary to prevent collapse. Thus, the more massive a star, the more luminous it had to be.

Eddington's theory implied that not all stars had to enter the Main Sequence at the upper left, as in Russell's theory. That would hold true only for the most massive of stars. A less massive star would never need to be so hot and luminous and could therefore enter the Main Sequence at virtually any point, depending on its mass. And rather than sliding downward along the Main Sequence curve, a stable star would spend about 99 percent of its lifetime at more or less the same point on the H-R diagram.

Meanwhile, Helmholtz's gravitational contraction hypothesis had fallen from favor, stumbling over the problem of angular momentum. Simply put, any body in motion around another body, or around its own center of mass, can be said to possess a supply of angular momentum. Angular momentum can neither be created nor destroyed, but it can be transferred from one body to another. In the case of Helmholtz's contracting nebula, some of the angular momentum in the spinning cloud would be imparted to the sloughed-off rings destined to become planets, but the major part should be retained in the Sun.

The Sun, however, spins very slowly (one revolution every 24.6 days at the equator) and, in fact, contains only about 2 percent of the solar system's angular momentum. Most of it—about 60 percent—belongs to the giant planet Jupiter, which has a rapid rotational period of about 10 hours. This state of affairs contradicted not only Helmholtz's theory, but also the entire Kant-Laplace nebular hypothesis.

If Helmholtz was wrong and gravitational contraction did not supply the Sun's energy, then what did? The glimmerings of an answer first appeared in the 1890s, when French physicist Antoine Henri Becquerel discovered the existence of radioactivity. Early in the twentieth century, British physicist Ernest Rutherford demonstrated the existence of the atomic nucleus, and Albert Einstein showed that a tiny amount of mass could be converted into an enormous amount of energy. Here was an obvious source for the Sun's energy: nuclear reactions changing mass into energy. The rate of conversion was such that the total loss of mass by the Sun was miniscule. It could go on releasing heat and light for billions of years before it finally exhausted its mass. The geologists and biologists got the time they needed, but the astronomers were left with very large unanswered questions.

What could account for the planets, if not the Kant-Laplace nebula? If the planets didn't form at the same time as the Sun, when did they form? And how?

Two American astronomers, Thomas Chrowder Chamberlin and Forest Ray Moulton, proposed a possible answer in 1906. They suggested that the Sun may have formed very much as described by Kant and Laplace, but without shedding rings of matter to form planets. The Sun would have existed alone, without its orbiting family. Then, several billion years ago, a maverick star invaded the Sun's territory. The Sun and the trespasser engaged in a gravitational tug-of-war, throwing off huge globs of stellar material. The maverick then proceeded on its way, and the liberated globs of matter cooled and became planets.

This theory seemed to resolve the question of where the planets came from, and it was virtually unchallenged throughout the first half of this century. One interesting implication of the theory was that life on Earth was, in all probability, unique in the universe. If planets form only as the result of interstellar traffic accidents, they must be extremely rare. One would expect only a handful of such collisions over the course of billions of years. For this reason, if no other, very few astronomers were willing seriously to entertain the notion of intelligent life's existing elsewhere in the universe.

When the collision theory was first proposed, it adequately explained the known facts about the history of the solar system; the problem was that very little was known. But during the first half of the twentieth century, the science of astrophysics came into its own. New technology—exemplified by the 200-inch Hale telescope at Mount Palomar, and by the later innovation of radio astronomy—and powerful new analytical tools—such as relativity and quantum theory—made it possible for scientists to study the heavens and attempt to explain, rather than merely describe, what they found there. By midcentury, the mass of new data about the universe forced scientists to reexamine and finally discard the collision theory. With many modifications, the nebular theory was revived.

The original nebular theory failed to account for the distribution of angular momentum in the solar system. The collision theory stumbled on the problem of planetary orbits. Any planets formed by a stellar fender bender would tend to have highly elliptical orbits, rather than the nearly circular orbits actually observed. Further, there was reason to doubt that hot globs of matter thrown off by a collision would ever be able to coalesce into planets. An acceptable new theory would have to address all of these questions.

By the 1940s, new versions of the solar nebula theory had begun to evolve. Swedish scientists Hannes Alfven and Svante August Arrhenius proposed a "renewable solar nebula model." In this scenario, the Sun forms first in the center of the nebula; its gravitational and magnetic fields then attract more dust and gas. The matter becomes ionized and is swept into circular orbits by the magnetic field of the Sun. Gradually, the small particles clump together, form their own fields, attract more matter, and slowly grow into planet-sized bodies. The process takes hundreds of millions of years.

A much shorter chronology was proposed by German physicist Carl von Weizsäcker in the 1940s. He analyzed the turbulence processes implicit in a body the size of the original solar nebula and concluded that such a flat, spinning disk would dissipate in a remarkably short period of time—hundreds to thousands of years, rather than millions. "This time scale," writes Harvard astrophysicist A. G. W. Cameron, "was so short as to be shocking."

The work of von Weizsäcker and others led to what is known as the "minimum solar nebula model." In this theory, the Sun and the solar nebula form independently, although possibly at the same time. The nebula contains only about 2 to 3 percent of the Sun's mass, but it is extremely efficient in forming planets. The turbulence in the nebula creates local areas of high density which, depending on the pressure and temperature in that region of the cloud, eventually condense into small, rocky planets or huge, gaseous ones.

The minimum solar nebula model works only if starbirth is assumed to be an incredibly efficient process. The competing "massive solar nebula model" grants nature a little more room to be sloppy. This theory supposes that the nebula contained two to three times the mass of the Sun. Most of the excess mass was lost early in the history of the solar system, leaving behind what we observe today.

It remains uncertain whether the solar nebula was minimum or massive, but there seems to be little doubt that the Sun and the planets did form from a collapsing, turbulent cloud of dust and gas. Theories are constantly evolving, as scientists focus on subtle details concerning pressure, temperature, composition, and the size and behavior of particles in the cloud. Although we have no final answers, we can, with some confidence, describe the life cycle of a typical star.

The Sun, 4.5 billion years old, still qualifies as a relatively young star. The universe itself was born between 10 billion and 20 billion years ago in an apocalyptic explosion known as "the Big Bang." For years, the Big Bang was thought to have happened more than 15 billion years ago, but scientists were forced to re-evaluate their theories in 1979, when it was discovered that some distant galaxies were moving twice as fast as had been thought. That meant that they reached their present locations in half the time that had been theorized; thus, the universe may only be about 10 billion years old.

THE LIFE CYCLE OF A STAR

The first generation of stars formed out of the debris from the Big Bang. These stars consisted almost entirely of hydrogen, but their fusion reactions created heavier elements. Over billions of years, these heavy elements were ejected into space and became the constituents of the gas and dust clouds out of which new generations of stars were formed. Today, star formation seems to take place exclusively in the arms of spiral galaxies, where there are ample supplies of dust and gas. In the older central cores of spiral galaxies, as well as in elliptical galaxies and globular clusters, there is little dust and gas and no evidence of on-going star formation.

The typical stellar nebula is a dark cloud about 10 parsecs across, with a temperature of about $10°K$ ($-263°C$, or $10°C$ above absolute zero) and a density of about 5,000 atoms per cubic centimeter. At some point, possibly spurred by external pressure from a nearby supernova, the cloud collapses and fragments into dense clumps, from which stars are formed.

Once the collapse has begun, a high-density core accumulates and the temperature rises. When the internal pressure reaches a high-enough level, the further inflow of mass is halted. This could account for what seems to be an upper limit on the size of stars. No stars with a mass greater than about 100 times that of the Sun are known.

A lower limit also exists. Clouds with too little mass can never generate temperatures high enough to touch off hydrogen fusion, the ultimate source of stellar energy. The planet Jupiter may be an example of such a case. Although it is huge in comparison with the Earth, Jupiter is not nearly massive enough to become a star; basically, it is just a large ball of hydrogen and helium. If Jupiter were several times more massive than it is, it might have become a dim companion star for the Sun. Perhaps as many as half of the stars in the Milky Way galaxy are binary- or multiple-star systems.

About 10,000 years after the cloud begins its collapse, a strong "solar wind" develops. Subatomic particles shoot outward from the core at very high velocities. After about 10,000 to 100,000 years, the excess cloud material has been dissipated, the solar wind dies down, and light from the "young stellar object" becomes visible. Thus, not counting the time it takes for the light to reach Earth, the youngest star we can observe is at least 10,000 years old.

By now, the "young stellar object" can be considered a star. Temperatures at the core reach $10^7°K$, hydrogen fusion begins, and the star enters the Main Sequence at a point determined by its mass. Over the next 10 billion years, the star will continue to burn hydrogen until about 10 percent of its mass has been consumed.

During this stage, a number of processes are at work inside the star. In order for a star to be stable, as Eddington pointed out, the internal pressure must balance the gravitational energy of the overlying mass. When instabilities develop, the star shrinks or expands in order to get back in balance. The fundamental relationship describing this process is known as the virial theorem. Where U equals the gravitational potential energy and K equals the internal kinetic energy, $2K + U = 0$. This means that half of the released gravitational energy is stored as heat, and the other half is lost from the system. In general, as energy is radiated away, the star shrinks to keep in balance.

In the core of the star, energy is thought to be produced by the fusion of hydrogen nuclei. Recent developments in physics have cast some troubling doubts on that long-held assumption. Although few, if any, scientists are ready to abandon the theory of hydrogen fusion in stellar cores, increasing attention has been paid to possible alternative mechanisms by which stars can produce energy. Some of those alternatives—and the reasons why they may be necessary—are considered in Chapter 4.

Hydrogen fusion proceeds via a process known as the proton-proton reaction. Two hydrogen nuclei (protons) combine to form a deuteron. The deuteron then captures a proton to form a helium-3 ($_2He^3$) nucleus. From there:

$$_2He^3 + {_2}He^3 \rightarrow {_2}He^4 + {_1}H^1 + {_1}H^1$$

The proton-proton reaction, in essence, turns four hydrogen nuclei into a single helium nucleus. But the atomic weight of hydrogen is 1.00813; that of helium, 4.00386. If we multiply 1.00813 by 4, we get 4.03252, which is greater than the weight of helium by 0.02866. In other words, we have about 1/141 of the original mass left over. This excess mass is the m in $E=mc^2$. It is what makes the Sun shine.

Although the proton-proton chain is the basic fusion reaction, it is not the only one. In more massive stars, where the core temperature may reach 18 million to 20 million degrees, energy may be produced by the carbon-nitrogen-oxygen (CNO) cycle. This is a more elaborate way of getting to the same place as the proton-proton reaction. Again, four hydrogen nuclei become a helium nucleus, but the CNO cycle includes a half-dozen intermediate steps. A single hydrogen nucleus combines with an atom of carbon-12 to form nitrogen-13. Hydrogen nuclei are then added successively to nitrogen-13, carbon-13, nitrogen-14, oxygen-15, and, finally, nitrogen-15. In the end, the carbon-12 "reappears," along with a helium nucleus and energy.

In still hotter stars, other elements may get involved in the reaction. After hydrogen, helium burning may begin, followed by carbon, neon, and oxygen. At the extreme temperature of 3 billion degrees, the star may burn silicon. By-products of the burning include elements as heavy as iron.

Iron is an extremely stable element, and the nuclear binding forces within the iron atom are exceptionally strong. Consequently, instead of fusing to form even heavier elements, iron reverses the process and breaks down to helium. Unlike other fusion reactions, the iron breakdown is energy *absorbing*. The necessary energy must come from the only available source, the gravitational potential energy of the star. This is released by the shrinking of the stellar core—an event which can lead to catastrophe.

Core shrinkage, in itself, is probably common in every star. As energy is released, the star must shrink somewhat to remain stable. When a star has exhausted about 10 percent of its total hydrogen supply, contraction begins in earnest. The core, by this time, is mostly helium, surrounded by a shell of hydrogen. With the fusion reaction proceeding in the hydrogen shell, the core continues to contract, becoming increasingly dense and massive. Finally, in compensation for the core shrinkage, the hydrogen shell expands. A deep convection zone develops, transporting more energy to the surface. The star becomes more luminous, although the surface temperature decreases. The expansion may continue until the star becomes a red giant.

At this point, the star finally moves away from the Main Sequence, toward the upper right of the H-R diagram. This region of the chart is populated by true "superstars," such as Betelgeuse and Epsilon Aurigae; the latter is more than 2 billion miles in diameter; if it switched places with the Sun, it would fill the solar system all the way out to the orbit of Uranus.

Stars in the red giant stage tend to be extremely cool at the surface; some have temperatures as low as 800°K. They may radiate most of their energy in the infrared wavelengths, making them difficult to detect optically. They are also incredibly tenuous. The interior of such a star is less dense than the Earth's atmosphere.

As a red giant consumes its remaining hydrogen, the helium core shrinks more rapidly. With contraction, the core temperature rises above that of the hydrogen shell. Helium begins to fuse, causing the core to expand while the hydrogen shell shrinks. The surface temperature rises, and the star moves to the left on the H-R diagram.

From this point, a star may go in any one of several directions. As the star shrinks, the deep convection zone may disappear, and the resultant drop in efficiency of energy transfer may cause a decline in luminosity. Or the star may begin to pulsate rhythmically.

By now, the helium in the core is depleted, replaced by carbon and oxygen. Surrounding the core there is now a double shell of hydrogen and helium, with most of the energy being provided by the hydrogen. But this is a very unstable situation, and the helium may flare up. In any case, as the mass of overlying hydrogen dwindles, the temperature drops and the helium-burning shell fades out.

When the total mass of the hydrogen envelope falls to about one tenth of a solar mass* the gravitational potential energy available is no longer great enough to balance the internal pressure. The envelope is violently ejected from the star, either in a series of spurts or in one great explosion.

What remains is a very hot, very dense core. The density may be so great that further contraction is impossible. The energy formerly held in by the hydrogen envelope now leaks away into space. The star may cool somewhat and enter the white dwarf stage, in the lower center of the H-R diagram. Here, stability returns, and the star may continue as a white dwarf for billions of years. Finally, its fuel exhausted, the star cools until it is no more than a dark cinder.

The end may be much more spectacular for a star that is 10 to 20 times as massive as the Sun. In massive stars, the core temperature may be much higher, so that when the helium core is exhausted, carbon burning begins. That may be followed by neon, oxygen, and silicon burning. Finally, the star has tapped every possible energy source except the last resort—gravitational energy.

* A "solar mass" is the amount of mass constant in our Sun. It is, in effect, a unit of measure; another star might have, for example, a total mass equal to 1.2 solar masses.

The core collapses upon itself, and a shower of neutrinos blows off the overlying matter in an explosion of inconceivable fury. The star has become a supernova.

In a matter of hours or days, a star that becomes a supernova may release as much energy as the Sun does in 10 billion years. For a brief period, a single supernova may be as luminous as an entire galaxy.

Supernovas occur in our galaxy at a rate of about three per thousand years. The last nearby supernova occurred in 1572 and was known as Tycho's Star, after the man who studied it in greatest detail. Tycho's Star was briefly bright enough to be visible in broad daylight.

What has turned out to be the most important supernova explosion took place in the year 1054 in the constellation Taurus. No European astronomer recorded the event (strictly speaking, in 1054 there were no European astronomers), but the Chinese kept careful note of the "guest star" and there is also evidence that American Indians portrayed the supernova in certain cave paintings.

The 1054 supernova is important because the debris from that cataclysm is still visible as the Crab Nebula. The Crab is a small patch of glowing gas that vaguely resembles its namesake but really looks more like a high-speed photograph of an explosion. The gas cloud is about 6 light-years across and is expanding at about 1,700 kilometers per second. Calculating backward in time, it is clear that the explosion must have occurred some 900 years ago; there is no doubt that the Crab consists of the remnants of the 1054 supernova.

In the 1950s, the new discipline of radio astronomy was responsible for the discovery that the Crab is a strong source of radio emissions. It releases about 10^{38} ergs per second, which is about 25,000 times more energy than the Sun emits. The source of all this power seems to be a rapidly spinning, pulsating object. It may, in fact, be a neutron star.

In a typical star, the density of the core is so great that matter becomes "degenerate"—that is, the atoms are so tightly packed that electrons are stripped away from their nuclei. In a neutron star, however, the density is so great that rather than being stripped away, the electrons are crushed inward toward the nuclei. The negatively charged electrons cancel out the positively charged protons in the nuclei, and the result is a dense mass of neutrons. Since angular momentum is conserved, the compact neutron star spins with incredible speed and emits regular pulses of energy; hence, neutron stars are also known as pulsars.

If a neutron star is massive enough, the total gravitational energy may be so great that nothing, not even light, can escape. It becomes a black hole.

Black holes may be the ultimate expressions of cosmological surrealism. These bizarre objects have inspired Disney movies, speculations that they may be interstellar transit stations, and philosophical tracts pondering the ultimate fate of the universe. Thanks to the work of the brilliant British physicist Stephen Hawking, we now have detailed mathematical descriptions of these strange phenomena, although, if history is any guide, we probably understand less about them than we think we do.

Our understanding of stars in general, though impressive, is still far from complete. Theories are in a near-constant state of revision as strange new phenomena pose even more questions. The most recent example is the 1979 discovery of a star dubbed SS-433, which displays a fantastic double Doppler. Most celestial objects have a Doppler shift toward the red, denoting motion away from the observer; a few objects have violet shifts, meaning they are moving toward us. But SS-433 has an extreme Doppler shift toward *both* red and violet. Superficially, this would mean that SS-433 is receding from us at about 20 percent of the speed of light—and approaching us at the same speed. The best explanation for this curious behavior seems to be that the star is rotating at a very high speed, and emitting huge jets of gas in opposite directions, rather like a Fourth of July pinwheel. The cumulative effect of the jets and the rapid rotation is to produce the double Doppler and the illusion of schizophrenic motion.

The discovery of SS-433 provides us with a much-needed caution. The universe is a place of bewildering diversity and complexity, and our understanding of it is—at best—limited. There are still many more questions than answers. From observations of literally billions of stars, we think we have gained at least a rough comprehension of the fundamental processes at work in the universe. But although we have seen the ends of the universe, we cannot see what lies just below the surface of a star only 93 million miles from home. The most puzzling star of all, in the end, is the closest one.

Two: THE VISIBLE SUN

In Tom Stoppard's brilliant play, *Rosencrantz and Guildenstern Are Dead*, the two Shakespearean spear-carriers at one point find themselves on a boat taking them to England and an uncertain but predestined fate. A confused and despairing Rosencrantz finally declares that he doesn't even believe in England. "Just a conspiracy of cartographers, you mean?" asks Guildenstern.

Many people share Rosencrantz's skepticism about the reality of places never seen. The interior of the Sun is one such place. Like England, it does exist, but there is a sense in which the solar interior is the product of a benign "conspiracy of scientists." We have never been there and we have never seen it, nor will we ever. Everything we think we know about the interior of the Sun is deduced or inferred from observations of the part of the Sun we *can* see. To extrapolate inward is a difficult business, particularly when the observable phenomena are so baffling in their own right. With the aid of physical laws, observation, computer modeling, and laboratory experimentation, scientists can construct highly detailed scenarios describing events in the solar interior. But in the end, our confidence in the reality of those events must depend on how well we understand the processes at work in the fraction of the Sun that is visible.

We return to the mysteries of the interior in Chapter 3; but first, a tour of the solar *exterior* is in order. The excursion will include four principal points of interest: the photosphere, the chromosphere, the transition zone, and the corona.

Solar energy begins with the fusion of hydrogen nuclei in the Sun's core, where the temperature may exceed 15,000,000°K. The energy created there takes tens of millions of years to penetrate the dense solar interior before reaching the visible surface, the photosphere, where the temperature is only about 6000°K. Through a series of complex interactions in the chromosphere, transition zone, and corona, the outer solar atmosphere is heated to a temperature as high as 3,000,000°K. Flares and prominences may lift huge quantities of superheated gases hundreds of thousands of miles above the surface. Beyond the low corona, the influence of the Sun's intense magnetic field declines, and subatomic particles escape into space, forming the solar wind.

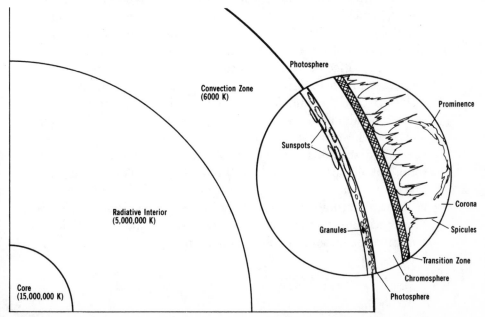

THE PHOTOSPHERE

To the naked eye, the photosphere appears to be the "surface" of the sun, although strictly speaking, a gaseous body like the Sun doesn't have a true surface. Nevertheless, the photosphere looks like a solid surface, and some early astronomers thought it was one. Sir William Herschel even supposed that beneath the shell of the photosphere, there was a cool region, "most probably . . . inhabited, like the rest of the planets."

The photosphere looks solid because it is much cooler than the solar interior. In effect, the temperature difference causes the photospheric gas to become foggy and opaque, giving the illusion of a sharply defined "edge" of the Sun. The mean temperature of the photosphere is only about 6,000°K, which is considerably cooler than what lies below. Surprisingly, the photosphere is also much cooler than what lies above it.

The opacity of the photosphere prevents observations of the solar interior. Whatever is going on inside is revealed only through its effect on the photosphere and the outer atmosphere. Fortunately, far from being the perfect, unblemished Aristotelian sphere it was once thought to be, the photosphere is actually a region of incredible turmoil and activity.

Sunspots are the most obvious expression of that turmoil. Chinese observers noticed them at least two thousand years ago. Western astronomers have kept track of them since the early seventeenth century. But if sunspots are easy to observe, they are also extremely difficult to understand. Writes Harvard solar physicist Robert W. Noyes, "We can't understand the first two things about them: why they exist at all and why they are black."

Galileo noticed the sunspots soon after he constructed his first telescope in 1610. He observed their motion across the solar disk, left to right, and found that a particular spot or group of spots took about 27 days to make a complete circuit of the Sun. This was the first evidence of solar rotation.

Later observers refined Galileo's data. Since the Sun is not solid, not all parts of it rotate at the same speed. The solar equator makes a complete rotation in slightly over 24½ days. At 60° latitude, one rotation takes nearly 31 days. This differential rotation is still not completely understood; it may be responsible for a number of important solar phenomena.

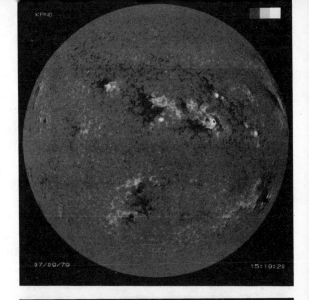

Astronomers conduct a daily Sun Watch, tracking the motions of spots, flares, and active regions across the slowly rotating disk. In this sequence, observed in hydrogen emission wavelengths, white and black areas indicate regions of opposite magnetic polarity. Note that most activity occurs in sunspot "zones" on both sides of the equator.

Kitt Peak National Observatory

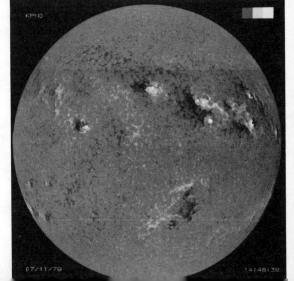

Galileo and his successors had no idea what the spots were. Some astronomers thought they were clouds, floating above the surface of the Sun. William Herschel took the opposite tack and proposed that the spots were actually breaks in the incandescent clouds of the photosphere. Others suggested that the spots were mountain peaks, storms, bubbles, meteorite splashes, or even slag heaps.

Whatever they were, the spots tended to appear in groups or pairs. They ranged in size from less than 100 miles in diameter to more than 100,000 miles. The largest sunspot group ever recorded was observed in April 1947. This mammoth group, visible to the naked eye, covered more than 6 million square miles, or about 1 percent of the visible solar disk.

The structure of individual sunspots is, as Gamow put it, "very complicated and grotesque." Typically, a sunspot consists of a dark central region, the umbra ("shadow"), encircled by a somewhat lighter ring, known as the penumbra ("partial shadow"). The penumbra is wispy and filamented and may appear to radiate from the center of the umbra; a round sunspot looks very much like the pupil and iris of a human eye. Most sunspots aren't round, however. Their shapes tend to be irregular and mottled, like clusters of inkblots. Within the irregular spots, the fine structure of the penumbra can be extremely complex—almost organic in appearance. The individual filaments, arched and sweeping, resemble Van Gogh's brushstrokes.

Spots and groups of spots endure for periods ranging from a few days to several months. Some have been tracked through several rotations of the Sun. In general, the spots grow in size and complexity before gradually fading away.

After two centuries of observing sunspots, scientists realized that there was a cyclical regularity in their behavior. At times of maximum sunspot activity, there may be more than a hundred sunspots visible; during slack periods there may be none at all for weeks at a time. After a 17-year study, in 1843 German astronomer Heinrich Schwabe identified what he believed to be a cycle of about 10 years separating periods of maximum sunspot activity. The figure was later refined to 11.2 years. In Switzerland, an amateur astronomer and pharmacist named Rudolf Wolf analyzed old astronomical records and found that the sunspot cycle seemed to hold up over a long time scale.

The existence of the 11-year sunspot cycle is an open invitation to those searching for correlations between events in the heavens and events on earth. The sunspot cycle has been tied to everything from the price of wheat and the Dow-Jones average to the incidence of political revolutions and the profits of the Hudson's Bay Company fur trade. I, myself, after about five minutes of diligent research, have just discovered a positive correlation between the sunspot cycle and stolen bases. The top ten performances by individual major league base stealers have all come in years of minimum or near-minimum sunspot activity. I don't conclude from this evidence, however, that Lou Brock, Ty Cobb, and Maury Wills all ran to the rhythm of the solar cycle. I conclude, rather, that coincidence is inevitable.

There have also been instances of apparent correlations between the sunspot cycle and terrestrial events which seemed to be real but ultimately turned out not to be. Around 1920, in the course of a study on tree-ring dating, American astronomer A. E. Douglass noticed that the growth rates of trees seemed to increase and decrease periodically in a 10- or 11-year cycle. This suggested a sunspot connection, but later analysis showed that the correlation was simply another coincidence.

Bogus correlations notwithstanding, there are some actual links between the sunspot cycle and earthly events. Magnetic storms and auroral displays tend to be larger and more frequent during the sunspot maximum. Radio operators have long been aware of the correlation between sunspot activity and the disruption of long-distance radio communications. The existence of such overt connections with the Sun has led scientists to speculate that more subtle connections may exist. There is some evidence to suggest that terrestrial weather and climate may in some way be tied to the solar cycle. The subject is still highly controversial and is discussed in some detail in chapters 7 and 8.

The reality of the 11-year sunspot cycle is well established. Its significance is another matter. There is some evidence that the 11-year cycle is not an essential, permanent feature of the Sun but, rather, a temporary phenomenon.

When Douglass was doing his tree-ring study, he found that the growth cycle was totally absent in wood samples from the late seventeenth century. This curious discovery led him to a paper written in the early twentieth century by E. W. Maunder, a

British astronomer. Maunder had undertaken a detailed study of
historical records of sunspot activity. His research revealed that
there had been a rather surprising hiatus in sunspot activity during
most of the seventeenth century.

Following the European discovery of sunspots in 1611, there
were two periods of maximum activity separated by about 15
years. By about 1645, sunspot activity had trailed off to almost
nothing. It remained at this extremely low point until 1715, when
the spots returned and settled into the familiar 11-year cycle.

Significantly, this sunspot holiday, now known as the Maunder
Minimum, occurred during a global era of low temperatures and
extended winters. In Europe and America, snow remained on the
ground as late as May or June, and the growing season was shorter
than normal. This "Little Ice Age," as it was called, began before
and ended after the Maunder Minimum, so it is difficult to demonstrate any causal links between the two episodes. Still, the coincidence is intriguing, and modern scientists are still attempting to
explain it.

Aside from their presumed effects on the earth, sunspots are
fascinating in their own right. After three hundred years of observing the spots, by the late nineteenth century scientists were at
last approaching a point where they could hope to explain them.
Earlier theories about mountains and meteorite splashes fell by
the wayside, victims of what NCAR solar expert John A. Eddy
refers to as "the limitations of astronomy without physics." The
vital clue to the true nature of sunspots was discovered, not by
an astronomer at his telescope, but by a physicist in his laboratory.

The dark Fraunhofer absorption lines in the solar spectrum
had been under intensive study for decades. By the 1890s several
of the more prominent lines had been identified as signatures of
elements present in the Sun, such as hydrogen and calcium (as
well as the bogus coronium). But as instrumentation improved
and resolution increased, astronomers found that the Fraunhofer
lines were multiplying alarmingly. Some of the lines could be identified as the products of different ionization states of the various
elements; at different temperatures, pressures, and densities, an
atom might lose one electron ("singly ionized"), two, three, or
many, with each condition resulting in a different spectral line.
When astronomers trained their spectroscopes on sunspots, however, they found that some familiar absorption lines were blurred
or even split into separate lines.

Dutch physicist Pieter Zeeman, working in his laboratory, found the explanation in 1892. He discovered that when a light source is placed in an intense magnetic field, its characteristic spectral lines are subject to the same sort of splitting and blurring that is found in sunspot spectra.

The cause of the Zeeman effect was not fully understood until the advent of quantum physics. Basically, the orbit of an electron moving in a magnetic field is altered according to the strength of the field. Altered orbits result in changes in the atoms' energy states, producing different spectral lines. The stronger the field, the greater the separation between the lines.

The Zeeman effect was discovered in a laboratory, but observations of nature in the raw are more difficult. American astronomer George Ellery Hale spent three years at the Mount Wilson Observatory trying to confirm the existence of the Zeeman effect in sunspot spectra. By 1908, he succeeded in confirming this, and, thus, the presence of strong magnetic fields in sunspots. The strength of the fields was a surprise. Earth's magnetic field measures about 0.3 gauss at the equator and 0.7 G near the poles. Hale found sunspot fields that measured 3,000 G.

The discovery of such strong magnetic fields in sunspots was like finding a solar Rosetta Stone. Suddenly, sunspots began to

The solar spectrum at a wavelength of around 5250 Å reveals absorption lines of singly ionized iron atoms. Some of the lines appear to have "doubled" due to the Zeeman effect, caused by the intense magnetic fields present in the Sun.

Kitt Peak National Observatory. Courtesy of Dr. John Harvey

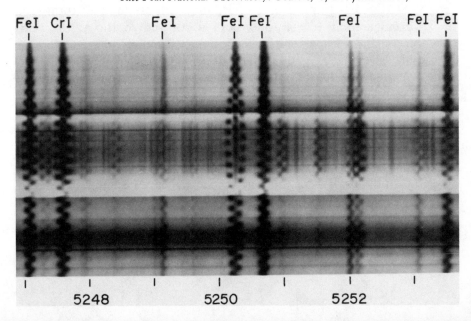

make sense—or at least, more sense than they had previously. The temperature of the sunspots had been found to be about 4,300°K, or 1,700°K cooler than the surrounding photosphere. This was consistent with sunspot magnetism, since magnetic fields tend to channel the normal flow of heat. The structure of the penumbra, with its long, arching filaments, was the result of the field's directing matter along magnetic lines of force. Even the tendency of sunspots to appear in pairs was a magnetic phenomenon; it was found that the paired spots possessed opposite magnetic polarity.

Consider a simple bar magnet, with two opposite poles, N and S. If you sprinkle iron filings around the magnet, the filings will arrange themselves, roughly speaking, along the magnetic lines of force. The same thing happens—speaking even more roughly— to the matter in the Sun. But there, the field is enormously stronger, and events are complicated by high temperatures and pressures.

The bipolar sunspot pairs seem to be the results of magnetic lines emerging from the Sun in one spot, then looping back in at the other spot. In a given hemisphere, the orientation of polarity is usually consistent, so that if the leading member of a pair is positive, the trailing spot is negative. In the opposite hemisphere, just the reverse is true. As a general rule, the hemispheres switch polarity every 11 years, and the magnetic orientation of the sunspot pairs is reversed. Thus, the 11-year sunspot cycle is really a 22-year cycle, encompassing two sunspot maximums and two polarity flips.

At the beginning of a cycle, the sunspots tend to form in a band around 40° latitude, both north and south. As the years pass, the sunspot zones slowly migrate toward the equator. By the eleventh year, they reach the equator and gradually die out. The polarity flips, and as the old spots fade, new groups form at 40°. These also move toward the equator, and after 11 years the polarity flips again, causing a return to the status quo at the end of 22 years.

The flipping of magnetic fields is not unique to the Sun. Earth's magnetic field also reverses itself periodically, but on a time scale of millions of years. No one yet knows why the Sun does it every 11 years, or why, during the Maunder Minimum, it apparently didn't do it at all.

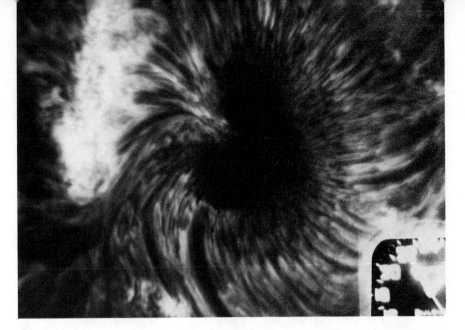

The magnetic nature of sunspots is revealed in this Earth-based picture. The field lines sweep the hot gases in graceful arcs (the penumbra) radiating from the central umbra. The temperature here is about 4300°K, some 1700°K cooler than the surrounding photosphere.
Big Bear Solar Observatory

The discovery of solar magnetism answered some questions, but it raised many others. Modern scientists still don't fully understand sunspots or their interaction with the rest of the Sun. At best, sunspots are simply one piece in an increasingly complex puzzle.

Another interesting piece of the puzzle is also found in the photosphere. When viewed through a powerful telescope, the photosphere takes on an appearance that one astronomer compared to a pot full of boiling rice. Known as granulation, this phenomenon seems to be an important link in the chain of energy transport from the solar interior. Photographically frozen in time, the network of granules vaguely resembles a honeycomb, with the bright, quasi-polygonal granules embedded in a web of narrow dark lanes.

The granules are hot blobs of gas, averaging about 1,800 kilometers in diameter. The lifetime of a typical granule is about 8 minutes from the time it rises to the top of the photosphere until it breaks up and descends in the dark lanes. The granules move vertically at a speed of about 0.5 kilometer per second. Except for the sunspots, the entire photosphere is covered with these seething gas bubbles; this apparently stable layer is actually continuously rising and falling through a space of about 25 kilometers.

The existence of granulation is an important clue to events beneath the visible surface of the Sun. The granules are thought to represent the top of a huge zone of convection which occupies the outer 15 percent of the Sun's radius. The hot gas bubbles rise from the interior to the cooler photosphere, where they release a small amount of energy before breaking up and descending. On the average, a granule cools by about 300°K while it is on the surface.

The granules contribute little to the overall temperature of the photosphere, but they seem to play an important role in the heating of the chromosphere and the corona. In addition, there have recently been hints that the entire Sun may be oscillating in a yet-to-be-explained 160-minute cycle. The granules may also be tied up in this oscillation, but no one can say precisely how.

Undoubtedly, there are structures in the photosphere smaller than the granules, which are about the size of Alaska. From earth-based observations, we can only see the photosphere to a resolution of around 200 kilometers. Even the proposed orbiting solar telescope will have a resolution limit of about 70 kilometers. Study of the small-scale structures in the photosphere will have to wait for close-up observations by unmanned space probes. NASA is currently planning at least two solar missions, one of which would send a spacecraft to within 2 million miles of the photosphere. A NASA study of the proposed mission concludes that, "Even a few images with ultra-high resolution would be immensely rewarding scientifically and would have an impact probably comparable to that of the first pictures of Martian craters or of the back side of the Moon."

THE CHROMOSPHERE

The photosphere, domain of sunspots and granules, marks the naked eye edge of the Sun. Above it, the solar gases are too tenuous to be visible against the white glare of the Sun's disk. But for brief moments during total eclipses, as the moon blots out the bulk of the Sun, we can get a glimpse of the rarefied outer atmosphere. It consists of the thin chromosphere and the billowing corona, separated by an abrupt transition zone.

The lower chromosphere is probably the coolest part of the Sun, with a mean temperature of about 4,300°K. Then, in the short space of about 2,000 kilometers, the temperature shoots up to around 50,000°K. This atmospheric heating is the reverse of what one might expect, and has been a continuing source of puzzlement for astronomers.

The elevated temperature of the chromosphere produces a spectrum different from that of the photosphere. In the cool photosphere, atoms steal tiny packets of energy from the interior, causing the dark absorption lines; in the chromosphere the atoms become excited and, rather than absorbing energy, release it. The result is a spectrum filled with bright emission lines in place of the dark absorption lines. The most prominent emission line is produced by hydrogen, and is known as the Hydrogen-alpha, or H-alpha line. It appears in the red part of the spectrum, thus giving the entire chromosphere a crimson tint.

Capping the chromosphere is an irregular boundary broken by jets of gas shooting up into the corona at 20 kilometers per second. These jets, called spicules, have inspired astronomers to grope for metaphors. The spicules, 5,000 to 15,000 kilometers high, rise and fall over a period of about 10 minutes, reminding Eddy of "choppy waves on a stormy sea." To nineteenth-century Italian astronomer Father Pietro Angelo Secchi, they looked like "a burning prairie." For Gamow, they resembled "a wheat field on a windy day."

Rising 5,000 to 15,000 km above the chromosphere, jets of superheated gases, known as spicules, define the boundaries of huge convection cells ("supergranules") in a pattern known as the chromospheric network.
Big Bear Solar Observatory

To examine the spicules and the chromosphere, astronomers use spectral filters, designed to screen out all but a specific wavelength of light. The resulting monochromatic images, known as filtergrams, provide a portrait of the Sun at a particular temperature. Since temperature varies with altitude, the filtergrams also serve, in effect, as topographic maps of the Sun. Typically, the chromosphere is observed through a filter keyed to the H-alpha wavelength.

Analyzing filtergrams of the chromosphere, astronomers found that the spicules were not simply random spurts of hot gas. Rather, they are arrayed over the solar disk in a fishnet pattern, known as the chromospheric network (see page 5 of color section). The spicules mark the boundaries of huge convection cells which are similar to the photospheric granules but on a much larger scale. The supergranules are about 30,000 kilometers across and each lasts for up to a day, as opposed to a few minutes for the photospheric granules.

Gas rises in the center of the supergranules, then spreads out toward the boundary spicules, and descends. This activity mirrors the motion of the granules, but is more regular and well-defined. Like the granules, the supergranules apparently have much to do with the heating of the corona.

The supergranules are associated with—and may cause—concentrated magnetic fields at the boundaries of the cells. The chromosphere, though much more tenuous than the photosphere, is still dense enough for the gas to dominate the magnetic fields. The convective motion of the gas shoves the field lines toward the supergranule boundaries, creating local areas of intense magnetism. Above the chromosphere, the gas pressure is less than the magnetic pressure, and the field lines assume a more regular shape. (See illustrations on page 3 of color section.)

Marking these regions of strong magnetic force, astronomers found features known as plages. Plages are areas of chromospheric brightening, visible in filtergrams. The plages tend to concentrate near sunspots; intriguingly, they generally appear before the spots themselves do and linger after the spots have faded.

Hot and dense, the plages are the spawning ground of solar flares. Flares are explosive eruptions of awesome power; even a modest flare would totally engulf the earth. Typically, a flare may last from 5 to 10 minutes, expending energy equal to the force of 10 million hydrogen bombs.

Flares generally occur in the vicinity of sunspots and are thus tied to the sunspot cycle. During the sunspot maximum, flares may go off once or twice an hour, producing the terrestrial effects usually blamed on the sunspots. A flare emits high-energy x-rays, ultraviolet light, and charged subatomic particles, all of which disrupt the ionosphere when they reach the earth.

The mechanisms involved in solar flares are bewildering and not very well understood. Robert W. Noyes describes flares as "perhaps the most complex phenomenon observed on the Sun . . . [they involve] an astounding variety of physical processes." Before attempting to describe those processes, it would be well for us to complete our "Cook's tour" of the visible Sun.

THE TRANSITION ZONE

Separating the upper chromosphere from the corona is the mysterious transition zone. This zone can be so thin as to be negligible; in other spots it can swell to a thickness of about 12,000 kilometers. The temperature in the transition zone ranges from about 150,000°K to 500,000°K.

The transition region is frequently described as "elusive," because it is extremely difficult to observe from the earth. Most of the emissions from this zone are in the form of ultraviolet light and x-rays, both of which tend to be absorbed by Earth's atmosphere. Short-lived sounding rockets provided the first detailed data on the transition zone, but it was not until the Skylab missions of the early 1970s that scientists were able to examine this region in depth.

In the transition zone, the well-defined chromospheric network begins to spread and blur; above the zone, the pattern of supergranules and spicules is all but indistinguishable. The transition zone thus seems to be the place where the density of gas declines to the point where it can no longer dominate the magnetic fields. Above the transition zone, as if in expression of the exuberance of liberation, the field lines become dominant and the temperature soars to 1.4 million°K.

The transition zone is not well understood, mainly because it marks the change from one set of processes (in the photosphere and chromosphere) that we barely understand to a totally different set of processes (in the corona) that we are only beginning to comprehend. If the transition zone were easier to observe, it might also be easier to understand; on the other hand, man has observed the photosphere daily for millions of years, yet we are still far from a complete accounting of its behavior.

THE CORONA

Until recently, the corona could not be observed at all, except for brief moments during total eclipses of the Sun. Solar eclipses, of course, occur when the Moon passes directly between the Sun and the Earth. The lunar shadow traces a narrow path across the surface of the Earth, no more than about 300 kilometers wide. An observer standing in the shadow sees the disk of the Sun gradually obscured by the Moon until, at totality, the photosphere is completely hidden. With the photospheric glare suddenly removed, the tenuous outer atmosphere of the Sun becomes visible.

The low corona is spectacularly visible for brief moments during a total solar eclipse. The tenuous gases of the corona are heated to more than 1,000,000°K and are swept outward along the lines of magnetic force generated by the Sun's intense magnetic field. The bright notch, at 8 o'clock in this picture, is an example of the eclipse phenomenon known as Bailey's Beads, caused by the reflection of sunlight through mountain passes on the limb of the lunar disk.

NASA

At any given location on Earth, the period between total solar eclipses can be measured in centuries. Since eclipses rarely come to astronomers, the astronomers are forced to chase the eclipses. This can be expensive and time-consuming. The total eclipse visible in the Pacific Northwest on February 26, 1979, was the last one North America will see in this century. The eclipse of February 16, 1980, was best viewed from central Africa and the Indian Ocean. July 31, 1981 will bring a glimpse of totality to the residents of Outer Mongolia. Stone Age tribes in New Guinea will have a fine view of the eclipse of June 11, 1983.

Eclipse watching can be a frustrating exercise. Totality lasts for an average of just 2 minutes, and no longer than 7. And eclipses, spectacular though they may be, enjoy no special meteorological privileges. Like baseball games and picnics, eclipses are frequently rained out.

It is not surprising, then, that modern scientists made no note of the corona until 1715. The corona is the thin, ragged outer atmosphere of the Sun. It may extend as far as 5 to 10 solar radii from the "surface" of the Sun. The corona consists of long, fan-like streamers radiating from the solar disk. Eddy compares their shape to that of tulip bulbs; to Gamow, the corona looked "very much like the hair of Albert Einstein in his old age."

Because the corona was visible only during a total eclipse, early astronomers were not even sure that it was a part of the Sun. Conceivably, it could have been evidence of a thin lunar atmosphere, or simply light scattered in Earth's own atmosphere. In fact, the light scattering occurs in the corona itself, as free electrons intercept light from the photosphere. However, the density of the outer corona is so low (10^{-16} grams per cubic centimeter) that very few electrons are available for light scattering. Consequently, the inner corona is about 100,000 times dimmer than the photosphere. The outer corona, beyond about five solar radii, is some 10 billion times dimmer than the solar disk.

Throughout the nineteenth century, astronomers gamely tracked down eclipses and tried to record the structure of the ephemeral corona. Drawings tended to be highly subjective, and not much was learned from them. Photographs weren't much better because of atmospheric glare and primitive equipment. American astronomer Charles Augustus Young managed to observe the spectrum of the corona in 1869 and found it was dominated by a strong green line which didn't correspond with any known lines in the solar spectrum; this was the birth of "coronium."

By the 1890s, astronomers at Lick Observatory in California were able to take the first good photographs of the corona by the use of a specially designed telescope. Nicknamed "Jumbo," the device was 12 meters long and had to be carted all over the world in pursuit of eclipses. Cumbersome but effective, Jumbo permitted the first truly detailed analyses of the corona and showed that it tended to be of maximum size and complexity during the peak of the sunspot cycle.

The body of data about the corona grew at an agonizingly slow pace. Waiting years and traveling thousands of miles for the sake of a few minutes of observation was clearly an inefficient way of attacking the problem. What was needed was a method of creating an artificial eclipse so that the corona could be studied in the comfort of an observatory. There were numerous attempts to build a device to do the job, but before 1930 all ended in failure. The problem was that the corona was so dim relative to the rest of the Sun that simple glare in the earth's atmosphere overwhelmed it.

Finally, in 1930 French astronomer Bernard Lyot succeeded in putting together a device that reduced glare and scatter light to an acceptable level. The Lyot coronagraph was effective only at high altitudes, above the dense air and effluvia of the lower atmosphere. Accordingly, Lyot lugged his invention to the top of Pic du Midi in the Pyrenees, 9,300 feet above sea level. There, he made the first non-eclipse observations of the solar corona.

Lyot's original model could see only the low corona, where the density was greatest. Even modern versions of the coronagraph are limited to observations out to about 2 solar radii (approximately 1 million miles). At best, coronagraphs are a poor substitute for an actual eclipse.

Despite the inherent limitations of the coronagraph, it provided the first detailed observations of the full spectrum of the corona. Since the corona is illuminated by light escaping from the photosphere, one might expect it to have a similar spectrum. But because the corona is so tenuous, it has a very weak spectrum with only a few faint emission lines.

The few emission lines that were present in the coronal spectrum failed to match the known elemental emission lines. Ever since Young's first glimpse of the green emission line in 1869, astronomers had been forced to assume the presence of hitherto unknown elements in the corona. This was a somewhat embarrassing position to be in, since by the 1930s, Mendeleev's periodic chart of the elements was getting crowded; there wasn't much room left for coronium. But data from the coronagraph finally permitted a solution to the problem. Analyzing the coronal spectrum, Swedish physicist Bengt Edlén found that the green line of coronium was actually a line emitted by highly ironized iron. Specifically, it was iron with thirteen of its twenty-six electrons missing.

The solution, inevitably, led to more puzzles. In order to get rid of half of the electrons of an iron atom, enormous temperatures are needed. The temperature of the corona therefore had to be in excess of 1 million°K. Since the photosphere was a mere 6,000°K, the question became: how does the corona get its heat? Today, scientists believe they have at least part of the answer, but the problem of coronal heating remains at the forefront of modern solar physics.

The structure of the corona presents less of a mystery. Early on, it was realized that the shape of the corona tended to follow magnetic lines of force. At the poles, in particular, the coronal streamers are arranged in patterns reminiscent of those formed by iron filings sprinkled over a simple bar magnet. The negatively charged electrons in the corona naturally cluster around the field lines, producing the sweeping arches and loops characteristic of magnetic fields.

Additional evidence for the dominance of the field in the corona comes from the structure of the solar prominences. The prominences, first observed during eclipses, are areas of dense material from the chromosphere that have been pushed upward into the corona and trapped there by the magnetic field lines. Prominences may endure for weeks or months and often form closed loops connecting back down to the chromosphere. Even a small prominence is much larger than the Earth, and many extend 100,000 miles or more above the photosphere.

The size of the corona is largely a matter of definition. Above 1 million kilometers, the influence of the magnetic field declines, and the trapped subatomic particles make a break to freedom.

The electrons, protons, and smaller nuclear debris stream outward from the Sun, extending the visible corona for several million kilometers. These particles make up the solar wind, which blows at a velocity of about 400 kilometers per second throughout the solar system. The influence of the solar wind is pervasive and its importance is only beginning to be appreciated. Undoubtedly it has an effect on earth, although the precise nature of that effect is highly controversial. From the Pioneer and Voyager missions to the outer planets, we know that the solar wind is strong enough to compress dramatically the magnetic fields of Jupiter and Saturn. Scientists believe that the solar wind extends the Sun's influence outward to at least 50 Astronomical Units (AU; that is, fifty times the distance from the Sun to the Earth) and perhaps even farther. Finally, at the heliopause, the outer boundary of the Sun's influence, the strength of the wind from our own Sun is overwhelmed by the combined wind from billions of other suns—a galactic wind.

Having taken a tour of the Sun all the way from its visible surface to the border of interstellar space, we can now plunge back into the unseen solar interior and attempt to answer the basic questions: how does the Sun produce its energy, and where does it go? Up to now, the tour has been like a cursory sightseeing excursion in Rome: we've seen the Colosseum, but not the catacombs. And it's in those shadowed recesses that the real secrets are buried.

Three: THE SOLAR BUDGET

Consider the subject of money for a moment. Dollar bills are printed at the mint, then shipped in bulk to banks throughout the land. Gradually, the bills make their way from the vaults to the tellers' cages and into the hands and pockets of the general public. From there, the money circulates from one cash register to another, back to the bank, then back out into the streets, over and over. Finally, dog-eared and ragged, the bills go from the bank back to the Treasury Department, where they are burned. Along the way, money frequently undergoes mysterious transformations. Coins become bills, bills become checks, checks become data bits in the innards of computers. And in the end, everything balances—theoretically.

Now consider energy as money (an easy-enough thing to do, times being what they are). In the Sun, energy is minted in the core, then gradually makes its way outward. Like money, the nuclear energy undergoes a variety of metamorphoses. It circulates back and forth, up and down, gets translated into foreign currencies, turns into small change, and ultimately, is lost from the solar economy. And theoretically, after all these transactions, everything balances.

Scientists attempting to balance the Sun's energy books have roughly the same task as economists trying to make sense of the Gross National Product. The economists must wade through the confusion of interest rates, exchange rates, discount rates, multiplier effects, and depreciation (not for nothing is economics known as "the dismal science"). For solar scientists, the accounting must include such items as thermal diffusion rates, internal rotation, convection cells, beta decay, pressure waves, and magnetic fields.

To make the energy books balance, scientists are forced to make assumptions about the nature of the solar interior without ever having seen it. The assumptions are constantly changing, as new data on the visible Sun forces refinements in the standard model of the interior. Recent developments have spurred a basic rethinking of the fundamental processes that may be at work in the Sun, but to date, nothing radically different from the textbook version has achieved general acceptance.

NUCLEAR REACTIONS IN THE CORE

The present theory of the solar interior began to take shape in the 1920s. The Helmholtz gravitational contraction hypothesis had fallen from favor, and scientists began to think in terms of nuclear reactions as the primary source of the Sun's energy. From Einstein's theories, as well as from the experimental work of Rutherford, James Chadwick, and others, it was known that some types of nuclear transformations liberated large amounts of energy. The problem was to pin down the precise reaction occurring in the solar interior.

In 1938, German-American physicist Hans Bethe proposed that the primary nuclear reaction in the Sun was the carbon-nitrogen-oxygen cycle. The most appealing thing about the CNO cycle was the fact that it *was* a cycle. The end product of the reaction was the conversion of hydrogen to helium, but the other elements involved were almost infinitely reusable. Also, the temperature required for the CNO reaction was about the same as the temperature of the solar core derived from Eddington's pressure/temperature/gravity equations.

At about the same time, Charles Critchfield suggested an alternative reaction, the proton-proton chain. Since both reactions ultimately led to the same result (helium plus energy), either or both seemed acceptable. As other scientists pursued the matter throughout the 1940s and 1950s, the proton-proton reaction gradually assumed the dominant position. The Sun was apparently not quite massive enough or hot enough to run on the CNO cycle.

By the end of the 1960s, it seemed fairly well established that the proton-proton reaction was responsible for about 98 percent of the energy produced in the Sun's core. The CNO cycle accounted for most of the remaining 2 percent, although other exotic variations of the fusion process seemed to be taking place on a small scale.

The core temperature was thought to be about 15 million°K. The density of matter in the core was calculated at 160 grams per cubic centimeter, compared with 10^{-9} g/cm^3 at the photosphere. Based on these calculations, the solar core ought to be about the size of the planet Jupiter, or approximately 75,000 to 100,000 miles in diameter. The force of gravity in the core is about 250 *billion* times as great as the surface gravity of the Earth.

In order for the energy books to balance, the core ought to be converting some 5 billion kilograms of hydrogen every second. If the reaction continued at this steady rate, the Sun could shine for 100 billion years. It won't, though, if present theories of stellar evolution are accurate. In about 5 billion years the hydrogen in the core should be exhausted, and the Sun will enter a phase of core contraction and surface expansion, and the Earth will be cooked to a cinder.

For now, however, the amount of energy reaching us from the solar core is, from a biological perspective, just about right. What most people don't realize, though, is that the solar heat reaching us today was created in the Sun's core tens of million of years ago.

The energy created by the fusion reactions in the core is in the form of high-energy gamma rays, at the short end of the electro-magnetic spectrum. Almost as soon as they are created, the gamma rays collide with other matter in the densely packed core. Because of the extreme temperature and pressure, matter in the core is "degenerate"; it consists of naked nuclei and free elec-trons, rather than discrete atoms and molecules. In this state, the opacity of matter is quite low, and the gamma rays would be able to escape with ease if it weren't for the high density. As it is, the gamma rays are continually being absorbed and re-emitted as they struggle outward from the core.

Eventually, after millions of years, the gamma rays emerge from the core and make their way through the thick radiative interior. Here, progress is even slower. Although the density of matter is less, the temperature is also lower. At lower tempera-tures, electrons are able to recombine with nuclei, forming photon-absorbing ions. The increased opacity of matter in this state makes it more difficult for the gamma rays to escape. With each absorp-tion and re-emission, energy is lost. By the time the gamma rays reach the surface of the Sun, most of them have lost so much energy from the repeated collisions that they are no longer gamma rays, but, rather, visible-light and ultraviolet rays. In effect, the gamma-ray dollar bills have been broken up into quarters, dimes, and nickels. Since energy is conserved, nothing has actually been lost, and the books still balance.

THE CONVECTION ZONE

Eighty-five percent of the way from the core to the surface, the process of radiative transfer breaks down. The temperature gra-dient between the hot interior and the cool surface is too great, the gas too opaque. Radiative transfer is effectively blocked. In its place, simple convection takes over. Convection is what hap-pens in a teapot on a hot stove. The water at the bottom of the pot is heated, while the water on top remains cool. But the cool water is denser than the hot water, and the situation is unstable. To reach equilibrium, the hot water rises to the surface, forcing the cool water down to the bottom of the pot; there, the cool, dense water is heated, and it then rises back to the top, replacing the original hot water which has by now lost some of its heat to the surrounding air.

In the Sun, the zone of convection occupies about 30 percent of the total solar volume. The density there is much less than in the core, so the huge convection zone contains only about 2 percent of the Sun's mass. As in the teapot, the superheated gases at the bottom of the convection zone rise toward the surface, appearing at the photosphere in the form of granules. The granules quickly cool and descend, replaced by other cells of hot, expanding gas. After many millions of years, the energy created by nuclear fusion in the core has finally reached the surface. Along the way, the temperature has dropped from 15 million to 6,000°K.

The overall up-and-down motion of the convection zone can be thought of as a wave. As any surfer knows, waves release energy. The waves propagated in the convection zone apparently reflect off the upper photosphere, creating what amounts to a continuous standing wave. When a wave propagated in a dense medium hits a less dense medium (in this case, the photosphere), it tends to be converted into a shock wave. These shock waves may penetrate as far as the upper atmosphere and release their energy there.

The existence of these waves complicates the aggregate energy budget of the Sun. It is rather as if our gamma ray dollars had been converted not just to nickels and dimes, but to pounds and francs as well. The total amount of wealth (energy) remains the same, but the differing exchange rates confuse the accounting.

The waves represent an enormous amount of energy, but unlike the radiative and convective energy, the wave energy can escape in a hurry. Just how much energy is in the waves and how quickly it is released is a matter of considerable controversy. Some recent evidence suggests that the waves may play a role in a suspected 160-minute oscillation of the entire Sun.

There is an age-old conflict in science between observers and theorists. Both groups are essential to progress, but each group tends to resent the other. Theories, by their very nature, are predictive, and the theorists expect the observers to find data to confirm their predictions. That happens frequently. But it also happens that observers find data that was not predicted and, in fact, has nothing to do with the theory. In such a case, the theorists often accuse the observers of sloppiness or worse. The observers tend to reply that the theorists don't want to be confused by facts. Historically, each side has had its innings. Many observers thought they saw canals on Mars; the theorists were certain that the canals couldn't exist as described, no matter what some half-frozen astronomer thought he saw through his telescope at three in the morning. The theorists were right; there are no canals on Mars. On the other hand, theorists spent hundreds of years trying to explain why Saturn was the only planet with rings. Those tightly reasoned theories all went down the drain in 1977 when observers found rings around Uranus; two years later, Voyager 1 discovered rings around Jupiter as well.

All of which is an elaborate preface to saying that the Sun may or may not be oscillating with a period of 52 minutes, or possibly 160 minutes. In 1976, while searching for something entirely different, Henry A. Hill of the University of Arizona found tentative evidence for the 52-minute oscillation. Meanwhile, three Soviet astronomers at the Crimean Astrophysical Observatory discovered a possible 160-minute cycle, unrelated to the Hill oscillations. The theorists, who had not predicted such oscillations and weren't really sure what to do with them, tended to doubt the validity of the observations.

The oscillation, if it exists, is small—only about 1 meter per second. This is close to the observational limit, which is one reason theorists are suspicious of the data. Another reason is that 160 minutes is precisely one-ninth of a day. It seemed highly unlikely that the Sun would be oscillating in harmony with the rotation of the Earth. The observers responded by refining their data, finding that the oscillation was actually 160.01 minutes, which is *not* a simple fraction of the Earth's rotational period.

The matter remains unresolved. According to Eddy, "This has to be considered one of *the* most controversial areas of solar physics. There are many people now who don't believe the Henry Hill oscillations . . . in part because you're dealing with a *very* small change." Small or not, the oscillations carry with them some very sizable implications. They would transport large amounts of energy from the interior very quickly, resulting in an embarrassing surplus in the solar accounting. To get rid of the surplus, the theorists would be forced to refigure the cash flow from some other important accounts, including nuclear fusion.

Hill's 52-minute cycle could be the result of pulsating pressure waves, or the so-called *p*-mode. If so, they would depend on the distribution of pressure, temperature, and density in the solar interior. They could provide information on the internal structure of the Sun in much the same way that earthquakes tell about the interior of the Earth. But the 160-minute oscillations would be much more difficult to explain solely in terms of the p-mode. They could conceivably be the result of gravity induced waves (g-mode), but there are serious theoretical problems here, too. In any case, the oscillations make it hard to come up with an interior structure that includes a core dense enough to maintain the proton-proton reaction. And without the proton-proton chain, the metaphorical solar marketplace would crash like Wall Street in 1929.

Whatever the role of the oscillations (if they truly exist), the wave action observed in the photosphere seems to be responsible for the elevated temperatures of the chromosphere and the corona. The energy requirements of the corona are enormous. In order to maintain the coronal temperature (between 800,000°K and 3,000,000°K), every square centimeter of the surface must release 10 million ergs per second, of which 1 million ergs must be deposited in the region of maximum temperature, some 50,000 kilometers above the surface. The bulk of this energy seems to be provided by the kinetic energy of the photospheric waves, granules, and spicules, although Noyes readily admits that the dynamics of the corona are "so complicated as to defy accurate prediction with our present understanding . . . the major physical processes are still poorly understood." The primary complication is introduced by the intense magnetic fields of the solar atmosphere.

THE SOLAR MAGNETIC FIELDS

The magnetic fields are clearly dominant in the corona. Matter is swept up by them, producing the characteristic looplike structures seen in the corona. Energy, too, is forced to flow along field lines, so that much of the energy released in the corona is effectively trapped there. The transition zone between the chromosphere and the corona seems to act as insulation, preventing the coronal heat from flowing back along the field lines down into the lower atmosphere. This reduces the overall energy requirements of the corona, but the interplay of forces in this region is so complex that no single mechanism—such as wave action or magnetism —seems adequate to explain all the observed phenomena.

Explaining and understanding the solar magnetic fields is crucial to any comprehensive model of how the Sun works, yet scientists are not even sure why planets and stars have magnetic fields in the first place. The whole question seems to be related to internal rotation rates, which cannot be measured directly. Until recently, it was considered significant that in every planet with a magnetic field, the magnetic axis is offset from the polar axis. On Earth, for example, the north magnetic pole is located in northern Canada, several hundred miles from the North Pole. But when Pioneer 11 reached Saturn, it found that the magnetic and polar axes of that giant planet are lined up almost perfectly. This discovery seems to cast some large doubts on the basic theory of planetary—and, presumably, stellar—magnetism.

Aside from explaining the origin of solar magnetism, scientists must also account for the 22-year cycle of polarity flips. In general, the interior of the Sun can be thought of as a reversing magnetic dynamo. The motion of hot gas across north-south field lines apparently creates currents which stretch the field lines to an east-west orientation. Convection then carries the lines of force upward, where the Coriolis forces of rotation turn them poleward again, but in an opposite orientation from the original north-south field.

If that is what is happening, there must be some mechanism that triggers the reversal in accord with the 22-year cycle. Attempts to construct computer models of the process are hampered by the manifold small-scale complexities of the Sun. It has been found, for example, that sunspots rotate faster than the gas they are embedded in. While gas at the equator takes 26 days to make a complete rotation around the solar axis, large sunspots on the equator make the trip in 25 days. Further, the rotation rates of surface layers appear to vary irregularly over the course of several months.

To make complete sense of the system, scientists must some-how connect the fields not only with the unseen interior, but also with the large-scale structure of the corona. That will not be easy, because the corona has turned out to be far more complicated than anyone suspected a few years ago.

THE COMPLEX CORONA

Even with the coronagraph, observation of the corona from the surface of the Earth is difficult. Early experiments with corona-

graphs in high-altitude aircraft and balloons returned disappointing results because the atmosphere, even at lofty elevations, is
still too bright. In the last two decades, better data has been
received from orbiting unmanned satellites. But it was the late,
lamented, much-maligned manned orbiting laboratory, Skylab,
that provided the first detailed look at the corona.

Skylab seems destined to be remembered more for its fiery demise than for its brilliant, productive lifetime. Between May 25,
1973, and February 8, 1974, three three-man astronaut crews spent
a total of 171 days aboard the laboratory. Although much was
learned about the long-term effects of weightlessness, the primary
scientific target of Skylab was the Sun, and it scored a bull's-eye.

Orbiting the Earth once every 93 minutes, 435 kilometers above
the surface, and equipped with sophisticated telescopes, cameras,
and filters, the Skylab astronauts could have an eclipse of the Sun
anytime they wanted one. Rather than settling for fragmentary
snapshots, the astronauts were able to observe the corona continuously for nearly nine months. The accumulation of new data was
nothing less than staggering. "So massive was the harvest of information," writes Leo Goldberg of the Kitt Peak National Observatory, "that it will be many years before the possibilities for productive analyses are exhausted."

Above the atmosphere, Skylab was able to observe the Sun in
wavelengths which were difficult to see from the ground. Especially important were the x-ray and ultraviolet observations, which
revealed for the first time the hidden structure and composition of the chromosphere, transition zone, and corona. From
ultraviolet (UV) spectra alone, sixty-five new emission lines were
identified.

Before Skylab, the corona could be seen only from the side, as
in eclipse observations. This limitation made it difficult to corre-
late events in the corona with events in the lower atmosphere.
But Skylab's x-ray instruments were able to see the corona face-on,
making possible a detailed, layer-by-layer analysis, from the outer
corona on down to the photosphere.

Even when solar activity is in the quiet phase of the cycle—as
it generally was during the Skylab missions—the corona contains
numerous active regions. These are large, hot, dense areas which
show up best in the x-ray wavelengths. The active regions generally appear above areas in the photosphere which are the seat of

strong bipolar magnetic fields. The familiar looplike structures connect spots of opposite polarity, forming magnetically closed structures. Associated with these bipolar regions are many small, bright x-ray sources.

Like sunspots, the active regions occur in well-defined latitude zones, known as activity belts. (See bottom picture on page 4 of color section.) They are symmetrical with respect to the equator, and seem to be the home of all sunspots and flares. The active areas are born, grow, and die, on a time scale of weeks to months.

The active regions are clearly controlled by the magnetic fields, but the complexity is such that it is difficult to understand precisely what is going on. The field lines may be twisted or laced together, like a plateful of spaghetti. Coronal arches connect active regions separated by many thousands of miles. Above the active regions, the coronal temperature is 0.5 million degrees higher than it is above quiet regions. The energy requirements of the active regions are thus much greater, and the source of the additional energy seems to be the magnetic fields. Magnetic force from the emerging active regions may somehow connect with pre-existing lines of force. In the process, the field relaxes and heat is released. However, standard theory holds that the reconnection should be accomplished by magnetic diffusion, and ought to take some 30,000 years. Using our money analogy again, the fields seem to function like slot machines, pouring out bonus energy. But this sort of slot machine should theoretically take 30,000 years to pay off; scientists are still trying to understand how the Sun beats the machine.

The key to the active regions—and, perhaps, to the entire Sun —may be contained in the small x-ray bright points. Earlier balloon and rocket observations had established the existence of many tiny x-ray sources, scattered over the entire globe of the Sun, but their significance was not appreciated until Skylab.

The bright points are linked with very intense, compact magnetic fields. Skylab astronauts watched them appear, brighten, and then fade on a very short time scale, ranging from about half an hour to eight hours. The bright points show a preference for the low-latitude activity belts, but unlike sunspots, they are not restricted to this region. Features similar to the x-ray bright points were also observed in the UV. (See top left photo, page 6 of color section.)

One of Skylab's most important and surprising discoveries was the fact that as overall solar activity declines, the population of bright points increases. This may reflect some internal balancing mechanism in the solar economy, keeping the total energy budget constant even during times of minimum sunspot and flare activity. Scientists are now coming to the view that the bright points may be a much more fundamental indicator of solar cycles than sunspots are. This opinion is reinforced by the fact that the total magnetic flux of the bright points is equal to or greater than that contained in the sunspots and active regions. If the bright points truly are the key, then, as Eddy points out, "in counting sunspots and plotting sunspot numbers, solar astronomers may long have been distracted." In effect, they may have been watching the holes instead of the doughnuts.

FLARES, PROMINENCES, TRANSIENTS, AND HOLES

The most spectacular of all solar phenomena is the flare. Scientists have long been aware that solar flares trigger magnetic storms and ionospheric disruptions on Earth, but before Skylab the origin of the flares was a nearly total mystery. They erupt unexpectedly in the chromosphere, "much like a pool of gasoline that has suddenly been ignited," as Eddy describes it. The flares release a deluge of x-rays and subatomic particles and liberate an astonishing amount of energy in just a few minutes.

Like so many other solar events, flares seem to originate in the magnetic fields. Hours to days before a flare erupts, excess energy is somehow stored in the magnetic fields, similar to the way energy is stored in the twisted rubber band of a toy airplane. One theory suggests that, as the pre-flare energy builds up, the field lines begin to pinch a suspected "neutral sheet," which separates oppositely directed lines of force. At some point, the neutral sheet is punctured, the field lines reconnect, and about half of the stored magnetic force is rapidly released as heat. The abrupt injection of energy blows away the overlying gases, creating the visible flare.

There are problems with this picture, however. Two unanswered questions are: how is the energy released so quickly, and why doesn't it happen sooner? The seeming contradiction in these questions only mirrors the inherent contradictions of the flares themselves. The release of so much energy (as much as 10^{32} ergs) in just a few hundred seconds is difficult to understand. On the other hand, what prevents a premature release of the energy? What keeps the rubber band from unwinding too soon? The role of the suspected neutral sheet is uncertain, because Skylab data indicated that the flares don't start down in the current sheets, but rather above them, at the tops of the magnetic loops.

Skylab x-ray pictures showed that the flares begin in concentrated magnetic loops, so small that even from orbit they are difficult to resolve. At the beginning of a flare, the tops of the loops brighten sequentially, like the tail lights of a Thunderbird or, as Eddy describes it, "like the popping off of a string of solar firecrackers." But these are no ladyfingers. The temperature in a flare may momentarily reach 20 million°K, 5 million degrees hotter than the solar core. In addition to the heat, flares eject subatomic particles at speeds approaching half the velocity of light. And unlike firecracker explosions, flare eruptions may last as long as several hours. Where the energy to maintain these long-lived flares come from is still uncertain.

If the flares are incredibly hot, one might expect them to play a central role in the solar economy. But Skylab revealed other events which, in their own way, are even more spectacular and perhaps more important than the flares.

Solar prominences had long been observed from Earth. They consist of dense blobs of chromospheric matter that get swept up in the magnetic fields and pushed into the corona. The prominences are relatively cool, averaging between 30,000°K and 90,000°K, and when seen in the H-alpha wavelengths, they occasionally vanish in just a few hours. Skylab x-ray observations showed that the disappearing prominences actually become superheated in the corona, making them indistinguishable at the lower H-alpha wavelength. By following the prominences into the upper temperatures, Skylab astronauts found them to be the triggering mechanism for the most extravagant solar display of all, the coronal transient.

Coronal transients are immense, sudden, almost surreal eruptions of the thin outer atmosphere of the Sun. In the course of just minutes, a transient billows outward, creating a mammoth bubble the size of the Sun itself. Before Skylab there had been some hints that such events might be occurring, but the detailed x-ray and ultraviolet observations from the orbiting laboratory provided a stunning confirmation of the reality of these truly awe-some phenomena.

Transients seem to be triggered by flares and prominences below the corona. When their energy is released in the corona, the coronal material is quickly accelerated to speeds as high as 1,200 kilometers per second. The electrons and nuclear debris in the corona shoot outward into space (or "the interplanetary medium," as scientists now prefer to call it), joining in the solar wind as it sweeps through the solar system. The transients appear to be the missing link between the flares and the Earth, and as such, they are of extreme interest.

Although the transients are enormous, they contain very little matter. They are, in fact, a much more perfect vacuum than can be created in a terrestrial laboratory. A trillion coronal transients, Eddy points out, would weigh less than the Earth. But compared with other regions of the corona, the transients are positively dense. It seems that the corona is riddled with holes.

Before Skylab, there had been some indications that the corona was not uniform over the entire Sun. Data from x-ray instruments on rockets revealed curious gaps in the coronal envelope. Intensive observations from Skylab confirmed the existence of the coronal holes and gave scientists a new insight into solar/terrestrial interactions.

The coronal holes are large areas where the magnetic field lines are open, rather than closed. The familiar loops and arches are absent, and x-ray and UV emissions are virtually nonexistent. The temperature in the holes is only about 1 million°K, or barely a third of the temperature in the active regions. The coronal holes are nearly always present above the solar poles, but they may also appear at lower latitudes. (See page 4 of color section.)

Beneath the holes, the transition zone is some three times as thick as at other locations. This swelling of the transition zone seems to reduce the loss of heat back into the lower atmosphere. In fact, there is no apparent connection between the holes and the structures of the photosphere and chromosphere, such as the granules and supergranules.

If the energy in the coronal holes is not flowing back down, yet not being emitted as x-rays or UV, where does it go? The answer is that some of it, surprisingly, is going to Earth.

Perhaps the most significant result of the Skylab missions was the discovery that the coronal holes are apparently the major source of the solar wind. Freed from the bondage of the field lines, charged particles such as electrons and protons escape into the interplanetary medium through the coronal holes. There is a strong temporal correlation between the appearance of such holes and the arrival of high-speed particles at our planet. There is considerable controversy over the meaning of it all, but at least some evidence exists that the particle streams may have an effect on terrestrial weather. The coronal holes have thus become the objects of intensive study.

Of course, only a miniscule percentage of the particles escaping the Sun ever hit the Earth, which is a very small target. In fact, most of the solar wind probably emanates from the long-lived coronal holes at the solar poles. Strong, continuous particle streams from the poles would direct much of the solar wind into the space above and below the plane of the ecliptic, which cannot be observed from the Earth.

Particles escape from all over the Sun, not just from the coronal holes, but elsewhere they result in more of a solar breeze than a solar wind. Trapped in the magnetic fields, the particles lose much of their energy before they can escape. But particles from the coronal holes retain most of their initial energy. Before the discovery of the holes, scientists had speculated about some sort of "nozzle effect" to account for the high-speed particles which were observed by early Soviet and American spacecraft. The coronal holes seem to fill the bill.

The escaping particles carry with them not just energy, but angular momentum. The original solar nebula hypothesis fell because it could not account for the Sun's apparent deficiency of angular momentum. But over the course of billions of years, the solar wind would dissipate a major fraction of the original angular momentum of the collapsing cloud. The wind would also have a braking effect on solar rotation; conceivably, the Sun may once have been rotating much faster than it does now.

In our inventory of the solar economy, the solar wind is roughly analogous to the petty cash fund. In itself, it doesn't amount to much, but juggling the petty cash account is frequently the only way to make all the books balance. The solar wind is the last link in the energy chain that begins in the Sun's core, but the final link is as important as any other. At virtually each stage of the energy-transport process, there are unresolved problems which involve, in one way or another, a surplus or a deficiency of energy release. A complete accounting of the solar wind may solve, or better define, some of those problems.

There remains one final, ironic twist to our energy-as-money metaphor. As any accountant, auditor, or FBI agent is well aware, even if the books balance, they may still be wrong. Fraud, counterfeiting, and embezzlement are facts of life, and there is a very real possibility that the solar bank has utterly bamboozled us. The only way to be sure is to look into the vault itself—the solar core. With an ingenuity that Willie Sutton would envy, in the last decade scientists have found a way to get around all the locks and seals and, at last, take a peek into the guarded inner sanctum of the Sun. What they saw there was more shocking than anything they could have imagined in their wildest nightmares.

The vault was empty.

Four: THE STRANGE CASE OF THE MISSING NEUTRINOS

Something is wrong with the Sun. Or, something is wrong with our understanding of the Sun.

Neither possibility gives much comfort to solar physicists. If something is truly wrong with the Sun itself—if somewhere in the deep solar interior the life-giving nuclear fires have been snuffed out—the implications for the Earth and all its inhabitants are grim indeed. If it is merely our understanding of the Sun that is faulty, then the rest of us can sleep easy, while the solar experts toss and turn, peer into shadows, and count, not sheep, but neutrinos.

Neutrinos are ephemeral scraps of nuclear debris, relativistic flotsam, hugging the thin edge of nonexistence. They have no mass except when moving at Einsteinian velocities. Their very existence was postulated in desperation, as a subatomic loophole in the laws of conservation of energy and angular momentum. They do exist, but they are almost impossible to detect. Nuclear theory insists that every time four protons fuse into a helium nucleus, two neutrinos must escape. The interior of the Sun should be the Detroit of neutrinos, mass-producing them, rolling them off the assembly line two at a time, endlessly. There should be more neutrinos than anyone could ever need.

But there aren't. Something has gone wrong.

The neutrino shortage worries nuclear physicists the way the gas shortage worries commuters. Without neutrinos, the great engine of physics sputters and coughs, and no quantum mechanic has yet been able to fix it.

"For astronomers," writes Harvard's Owen Gingrich, "this is the real solar energy crisis." Gingrich's use of the word "crisis" is not simply hyperbole. His colleague Robert W. Noyes describes the neutrino problem as "a crisis of confidence in basic theories of stellar structure and evolution . . . it is threatening to topple the entire structure of stellar interior theory." UCLA astronomer Roger K. Ulrich maintains that until the neutrino situation is resolved, "we cannot have confidence in any predictions based on the theory of the solar interior."

As early as the 1930s, knowledge of the solar interior seemed to be, as Noyes put it, "well tidied up." Scientists were already looking around for more challenging problems. But at the end of the 1960s, suddenly and surprisingly, scientists found themselves being critically zapped by the neutrinos. Throughout the 1970s, solar physicists went about their work with the look of a man who had just discovered that he built his dream house on top of the San Andreas Fault.

NEUTRINO THEORY

Neutrinos were born in controversy in 1933. Physicists working with the first particle accelerators found a troubling anomaly in the radioactive decay of certain atomic nuclei. In a process known as beta decay, a nucleus ejects an electron and releases energy.

Theoretically, the energy should have gone into accelerating the electron, but that was not happening in every case. Typically, the electron absorbed only about half of the released energy. What happened to the rest of the energy was a mystery.

Rather than abandon one of physics's most fundamental tenets, the conservation of energy, Austrian physicist Wolfgang Pauli suggested that missing energy from beta decay was being carried away by an undetected subatomic particle. The hypothetical particle would have no charge and no mass. The idea of an object with no mass is difficult to comprehend, but in the wonderland of quantum theory, it makes sense. In quantum theory, particles don't exist, they merely *tend* to exist, at preferred locations and energy levels. Neutrinos tend not to exist at rest (i.e., they have zero "rest mass"); thus they have no mass except for what they acquire by acceleration to relativistic velocities.

Later generations of particle accelerators confirmed the existence of Pauli's hypothetical neutrino. It turned out to be merely one of dozens of subatomic particles spewed out by nuclei under the extreme conditions created in modern atom smashers. The neutrino's primary distinction is its near-total disdain for other matter. Possessing no electrical charge, it is not subject to the influence of charged particles such as protons and electrons. Having no mass, it seldom collides with anything. A neutrino could pass unscathed through a wall of lead a thousand light-years thick.

There are no known lead walls of such dimensions, but the interior of the Sun is a formidable obstacle in its own right. We have seen how gamma particles must battle for some 50 million years to escape the crowded solar interior. But neutrinos don't even notice the congestion; they escape instantly.

Scientists realized that neutrinos offered the only hope of "seeing" the solar interior. Radiation from the surface of the Sun is old news, delivered on foot. Studying it is useful, but it's like going to the mailbox and finding a copy of a 1927 *Time* magazine, trumpeting the news that Lindbergh made it to Paris. That's certainly interesting news, but if it is our only source, then we'll have to wait years to find out about subsequent events in Paris, such as the German occupation and the current height of hemlines. Neutrinos would give us, in effect, a radio correspondent on the scene, broadcasting the news as it happens.

No one really expected that the news would be very startling.

Basically, it was thought that a neutrino-detection experiment would serve as a useful check on the validity of standard theories of stellar evolution and nuclear fusion. No one doubted that the solar neutrinos existed; the problem was to find them.

In the mid-1960s, Raymond Davis, Jr., of the Brookhaven National Laboratory, devised an elaborate and seemingly foolproof method of trapping the solar neutrinos. The vast majority of neutrinos should be produced by the basic proton-proton reaction, but these neutrinos would have an energy level too low to permit easy detection on Earth. There is, however, an alternative path which is sometimes followed by helium-3 nuclei in the course of nuclear fusion. Occasionally, such a nucleus may fuse with a helium-4 nucleus to produce a nucleus of beryllium-7. The beryllium-7 nucleus then collides with a proton and becomes boron-8. The boron-8 nucleus is unstable and quickly decays into beryllium-8, which then splits into two helium-4 nuclei, returning to the end product of the basic proton-proton chain. This side branch of the reaction has no effect on energy production, but, as boron-8 decays, it emits a very high-energy neutrino. This was the neutrino Davis planned to capture.

THE HOMESTAKE EXPERIMENT

Most astronomers build their observatories on top of mountains. Davis built his at the bottom of a gold mine. Most scientists probe the universe with the aid of complex, sophisticated instruments. Davis's neutrino-detection instrument was, in essence, simply a vat of cleaning fluid. But if the Davis experiment was prosaic in its physical circumstances, it was nevertheless marvelously ingenious in design and concept.

Calculations showed that a neutrino emitted by boron-8 decay should have enough energy so that if it collided with an atom of chlorine-37, the target atom would be transmuted into argon-37. The process is simply the reverse of beta decay. Neutrino collisions with other atoms may produce similar effects, but for a variety of reasons the chlorine-argon reaction was seen as the best bet for neutrino detection. Chlorine is cheap and abundant, an important consideration for scientists in search of funding; and argon is an inert gas which refuses to combine with other elements, thus making the collision product easy to detect and measure.

The Davis experiment consisted of a 400,000-liter tank filled with the cleaning fluid perchloroethylene (C_2Cl_4). In order to shield the chlorine solution from random reactions with cosmic rays, the tank was placed in a specially prepared chamber a mile underground, in the Homestake Gold Mine in Lead, South Dakota. Homestake is an active mine—in fact, the most productive in the Western Hemisphere—and the technique used by Davis was not unlike that of the gold miners. In effect, he was searching for nuggets of argon in the midst of the chlorine ore.

The boron-decay branch of the proton-proton chain is rare in comparison with the usual reaction, but Davis and his colleagues were confident that there were enough of the high-energy neutrinos around to yield a significant harvest of argon. The standard models of the solar interior predicted that 5.6×10^{-36} neutrinos would be captured per second per target atom. That means that a given atom would only be hit once in about 10^{30} years, which is about 10^{20} times the age of the universe. But there are a lot of chlorine atoms in 400,000 liters. During a 100-day test run, the tank should have collected about fifty atoms of argon-37.

The Brookhaven Solar Neutrino Experiment is located 4,900 feet beneath Lead, South Dakota, at the bottom of a gold mine. The tank, 20 feet in diameter and 48 feet long, contains 100,000 gallons of the cleaning fluid perchloroethylene, whose chlorine atoms may capture solar neutrinos. Initial results from the experiment produced a crisis in solar theory; the number of neutrino captures was far below the standard predictions.

Brookhaven National Laboratory

To simplify things, a capture rate of 1.0×10^{-36} captures per second per target atom was declared to be 1 Solar Neutrino Unit, or SNU. Of the 5.6 SNUs predicted for the experiment, 4.3 would come from the boron-8 neutrinos. The remaining 1.3 would be produced by other branches of the fusion reaction. That, at least, was the prediction.

Davis and his Brookhaven colleagues began running their experiment in the early 1970s. In an article in *Science*, Davis and his co-experimenter John N. Bahcall, of Princeton's Institute for Advanced Study, reported their findings: "All of us have been surprised by the results; there is a large, unexplained disagreement between observation and supposedly well-established theory." Understatement is the accepted scientific style in these matters. If a team of medical experimenters ran a routine test to determine the amount of blood in the human body and found not a single drop, their report would probably begin, "All of us have been surprised. . . ."

All of them *were* surprised—profoundly. The expected argon production rate should have been about 1.1 atoms per day. The observed production rate was 0.13 atoms per day—plus or minus 0.13. Since the cosmic-ray background, even a mile underground, should have produced 0.09 ± 0.03 argon atoms per day, the experimental results were indistinguishable from the background. Simply put, the neutrino detector had failed to detect any neutrinos.

The 5.6 SNU prediction was blown out of the water. The initial results from the Brookhaven experiment put an upper limit of about 1.0 SNU on solar neutrino production, and even that was optimistic. Since the 5.6 SNU prediction was based on a model, a result a little higher or a little lower would not have been terribly significant. But a result of 1 SNU not only destroyed the standard models, it actually challenged the very concept of nuclear fusion as the source of solar energy.

THE SEARCH FOR ALTERNATIVES

When an experiment turns up a totally unexpected and startling result, the first instinct of scientists is to doubt the validity of the experiment. This is especially true in a case in which the results undermine fundamental theory. When the news from Homestake was announced, scientists all over the world examined the details of Davis's experiment and tried to find a flaw. This was not petty jealousy; if there was something wrong with the experiment itself, Davis and his colleagues were as eager as anyone else to find out what it was.

Most of the attention centered on the argon retrieval process, since that was really the only part of the experiment where anything *could* go wrong. As a test, Davis introduced 500 atoms of argon-37 into the tank, and then attempted to retrieve them by the same method used to measure the neutrino-produced argon. The recovery rate was about 90 percent, more than enough to support the validity of the experimental technique.

If the experiment was working as predicted, then the flaw had to lie elsewhere. Speaking broadly, there were only four other possibilities: The calculated rate of neutrino/chlorine interactions (known as capture cross sections) could be wrong. A second, more troubling possibility was that the standard models of the solar interior were faulty. An even more troubling possibility was that the understanding of basic physics was somehow inadequate. Or, worst of all, the Sun itself could be malfunctioning.

It is fair to say that most scientists would have been greatly relieved if the fault had been found to reside in the neutrino cross-section calculations; such a finding would have the minimum impact on solar physics. But one of the reasons chlorine was chosen for the experiment in the first place was the fact that its cross sections were already well established. Work in a number of laboratories, according to Davis and Bahcall, had provided "a solid experimental basis for the original theoretical calculations of the neutrino capture cross sections of ^{37}Cl." Measurements made at the Kellogg Laboratory of the California Institute of Technology, and confirmed at other laboratories, supported the cross-section predictions used in the experiment.

If the cross sections were accurate, then the standard solar models were suspect. Even "nonstandard" models were in trouble. Ulrich noted that "the smallest rate of neutrino production which has been achieved in a nonstandard solar model is 1.4 SNUs." That was still too high.

One can play games with numbers. The standard models gave a solar core temperature of 15 million°K. If the temperature were 1 million degrees lower, the helium-3 nuclei would not be able to combine with helium-4 to produce the side branch of the reaction responsible for boron-decay neutrinos. That would eliminate 4.3 SNUs, but leaves 1.3 SNUs that should have been observed but weren't.

Scientists began to speculate about processes which could cause core reactions to switch on and off. A lower core temperature might make the proton-proton chain less efficient, resulting in a build-up of helium-3. At some point, the excess helium-3 mixes into the core and burns rapidly. This sudden release of energy would cause the overlying layers of the interior to expand, which, in turn, would cool the core and turn off the neutrino-producing reactions.

The problem with this model and with other similar models is that they demand periodic variations in the luminosity of the Sun. The helium-mixing model, for example, would lead to a 10 percent variation in the Sun's luminosity over the last million years. The effects of such a change would hardly go unnoticed on Earth. Average terrestrial temperatures should change by about $2°K$ for every 1 percent change in solar luminosity. Yet there is no evidence that global temperatures have changed by $20°K$ over a time span as short as a million years.

Even longer-range variations present difficulties. The basic limitation on solar variation is defined by our own existence. If the Sun ever gets much hotter or cooler than its present temperature, life on Earth would be impossible. Planetary scientists have calculated that if, at some point in the past, the Sun had been 10 percent less luminous than the present value, Earth's oceans would have frozen, never to remelt. Although some scientists challenge that conclusion, it is nevertheless clear that any major variations in solar luminosity would have devastating effects on the Earth.

The solar model makers are thus confronted with strict limitations on the magnitude of possible changes in the Sun's core. Whatever is happening in the core must ultimately be reflected in the surface luminosity of the Sun and the global temperature of the Earth. Depressing the core temperature enough to get rid of the neutrinos seems to lead inevitably to a frozen Earth. Yet maintaining terrestrial temperatures in a biologically comfortable range seems to require an energy output from the Sun that demands neutrino production. In addition, the models must describe a Sun which behaves like a proper Main Sequence star; no special pleading is permitted.

As theorists struggled to explain the neutrino shortage, the word "crisis" began to appear in the scientific journals. The neutrino problem had a ripple effect. If scientists were so badly wrong about the Sun, what did that say about their understanding of stellar evolution? And since virtually the entire structure of astrophysics rests on the present theories of stellar evolution, might the entire house of cards come tumbling down? Is everything we know wrong? Could it be that twentieth-century science is the result of a monumental hubris? Are we, despite our supposed sophistication, as naive about the universe as the Greeks and Egyptians? These are not idle, philosophical musings. The neutrino problem has made them very real and concrete questions.

The Homestake results came uncomfortably close, as Ulrich put it "to ruling out the assumption that nuclear reactions are currently producing the luminosity of the Sun." But if nuclear fusion was not doing the job, then what is?

In 1975, M. J. Newman and R. J. Talbot suggested a possible alternative. What if there were a black hole in the center of the Sun? Newman and Talbot calculated that such a black hole could have a mass equal to about 1/100,000 of the total solar mass. The hole's radius would be no more than a few centimeters. As mass fell into the hole, it would radiate enough energy to account for about half of the total solar luminosity. If the remaining half were provided by the proton-proton reaction, the observed neutrino flux would have been around 1 SNU, which closely matched the early Brookhaven results.

This is an extreme suggestion, and most scientists are far from ready to accept it. Yet, as Noyes writes, "In what some might classify as desperation, other even more bizarre suggestions have been made to resolve the neutrino problem." Among the most bizarre is British cosmologist Fred Hoyle's theory that the origin of the Sun was considerably different than the standard theory proposes. Hoyle has long had a reputation as a scientific gadfly, ever ready to defend discarded hypotheses and propose strange new ones. But Hoyle's scientific credentials are sound, and his radical theories are not easily disproved. In this case, Hoyle suggests that the Sun formed in a two-stage process. The inner core, with about half of the present solar mass, would have formed much earlier than the rest of the solar system. Orthodox theory holds that the entire Sun formed some 4.6 billion years ago, but, according to Hoyle, that was merely the time at which the early core accreted the surrounding outer layers that make up the present-day Sun. The difference in composition of the two halves would produce a highly convective core, and a resultant drop in neutrino production.

Ranging even further afield, one intriguing theory by Robert Dicke of Princeton suggested that the answer to the problem may lie in orbital perturbations of the planet Mercury. Again, this was no crackpot hypothesis. Dicke proposed a possible extension of Einstein's theory of relativity, known as the "scalar-tensor" theory. One of the proofs of relativity came from observations of the orbit of Mercury. Old-style Newtonian-Keplerian physics could not account for certain perturbations in Mercury's orbit. Calculations which allowed for relativistic effects produced by the proximity of the Sun produced a much better explanation of Mercury's motion. But Dicke's scalar-tensor theory suggested that relativity only accounted for 93 percent of the observed perturbations.

The remaining Mercurian wobble might be explained, Dicke suggested, if the Sun were slightly oblate. If the equatorial radius of the Sun were about 30 kilometers greater than the polar radius, the solar magnetic field would take on a high order complexity which would affect the orbit of Mercury. To get an oblate Sun, Dicke suggested that the solar interior could be rotating as much as twenty times faster than the surface. As a kind of by-product, the rapidly rotating core would decrease the internal temperatures and pressures enough to reduce the neutrino flux to an acceptable level.

Dicke's theory could be proved by direct observation, but it wasn't substantiated. With an instrument built specially for the task, Henry Hill and his colleagues looked for Dicke's oblateness in 1975 and didn't find it. What Hill *did* find, though, was the 52-minute oscillation discussed in Chapter 3. That led to questions about the density of the core and its ability to support the proton-proton reaction . . . bringing us right back to square one.

By the late 1970s, the situation was a little less grim. The Brookhaven neutrino experiment was statistical in nature, and, as in any statistical process, the more data, the better the accuracy. Continued experimentation at Homestake revealed an upward trend in the number of neutrino captures. At first the experimenters suspected that this was simply a statistical fluctuation, but the trend held up over a long time scale. From an initial result of 1.0 SNU, the accepted figure rose to 1.4 by 1976, and by the beginning of 1980 it stood at 1.7 ± 0.4 SNU. This was still far below the 5.6 prediction, but it was high enough to ameliorate some of the more extreme consequences of the experiment. In particular, the concept of nuclear fusion in the solar core seems to have been rescued from limbo.

But the problem is still present, with no obvious solution in sight. Astronomers and physicists bat the question back and forth like a shuttlecock, with neither group wanting it to land in its court. Davis and Bahcall neatly summarized the chaos they had created: "The attitude of many physicists toward the present discrepancy is that astronomers never really understand astronomical systems as well as they think they do, and the failure of the standard theory in this simple case just proves that physicists are correct in being skeptical of the astronomers' claims. Many astronomers believe, on the other hand, that the present conflict between theory and observation is so large and elementary that it must be due to an error in the basic physics, not in our astrophysical understanding of stellar evolution."

With the astronomers looking over their shoulders, the physicists have been forced to do some rethinking of their own. Nuclear theory is riding high these days, thanks to remarkable agreement between theory and observation. Physicists believe they are getting close to achieving the goal that eluded Einstein—a unified field theory, in which all of nature's fundamental forces are neatly linked to one another.

But as the physicists squeeze more and more basic particles out of the cosmos (including strange beasts with names like "naked charm," and "bare bottom,")*, the now mundane neutrino is

---

* Current theory postulates a family of fundamental particles called quarks, after a line in James Joyce's *Finnegans Wake*. The names of the individual quarks are equally whimsical. Among the proposed quarks are "strange," "charm," "truth," and "beauty," or, alternatively, "top" and "bottom." Thus, one might find quarks in combinations that are strangely beautiful; alone, they might be topless or bottomless. Science marches on.

demanding a closer look. Perhaps in the current controversy, the fault lies not in our star, but in our neutrinos.

The Sun ought to be running on nuclear fusion, and if that is the case, it is difficult to avoid producing neutrinos. But it is 93 million miles from the Sun to the Earth, plus 1 more mile to the bottom of the Homestake mine. Could something be happening to the neutrinos along the way? Bahcall has suggested that the stable neutrinos observed traveling distances of less than 1,000 centimeters in laboratories might somehow prove unstable over the 10^{13} centimeters between Sun and Earth. If the neutrino possessed a rest mass that was not dead zero, then conceivably it could decay into some (undiscovered) particle during its passage to Earth.

Such a solution would go a long way toward getting the solar astronomers off the hook, but it would not be appreciated by the physicists. Not surprisingly, the suggestion of neutrino decay has not been well received. For one thing, no evidence has been found to support the idea.* That pleases the physicists, since nuclear theory works much better if the rest mass of neutrinos remains at zero.

---

* This may no longer be true. As this book goes to press, physicists are rethinking the entire neutrino question. In May of 1980, Dr. Frederick Reines and his colleagues at the University of California at Irvine announced new experimental results which seem to imply a non-zero rest mass for the neutrino. Reines found that neutrinos apparently oscillate between distinct states ("flavors"), and the existence of these oscillations seems to demand a small and as yet unmeasured rest mass. This finding has some interesting implications. The solar neutrinos may not be missing, but merely "hiding" from our detectors by way of the oscillation; thus, the Sun may be okay after all. Another spin-off of the discovery may be an answer to an enduring cosmological question. The universe is known to be expanding, but that expansion could be halted by gravity if the total mass of the universe is great enough. If neutrinos do have mass, then there is reason to think that the expansion will ultimately turn to contraction. The universe would thus oscillate between expansion and contraction, and each cycle would be kicked off by a Big Bang. Our universe may be merely one in an infinite series of universes. Or not. At this point, no one is really sure what the Reines experiment means. "The universe," says Reines, "is not the way we thought." Just what way the universe is remains to be determined, but the oscillating neutrinos may turn out to be the key. Stay tuned.

If the problem cannot be deferred to the physicists, nor solved by the astronomers, nor blamed on the experimenters, then the Sun itself is the only remaining suspect. Although any theory which makes the Sun an abnormal star has been frowned upon since the time of Copernicus, the possibility remains—perhaps something extraordinary has happened to the Sun.

Solar models which cause significant variations in luminosity inevitably run aground on the shoals of the Earth itself. If the solar core periodically switches itself on and off, there ought to be abundant paleoclimatic evidence on Earth. One immediately thinks of ice ages. But there are other explanations for ice ages which do not invoke changes in the Sun. There is some evidence to support minor changes in solar luminosity (see Chapter 6), but those changes are probably not large enough to account for a total shutdown of neutrino production.

So it is at least possible—it is far too early to say probable— that something is happening in the solar core that has not happened before. The effects of the core event are not yet manifest in the surface luminosity; but sometime in the next 30 million to 50 million years, they may be. When that day comes, if it ever does, life on Earth could become precarious. Perhaps, as oceans freeze and glaciers advance, our remote descendants will pack up and move to a terra-formed Venus, closer to the dying solar hearth. Perhaps they will leave the solar system entirely, in search of a more dependable star. Or perhaps they will just bundle up and wait for the return of the good old days, when the rising Sun brought warmth to a green and growing Earth.

It is still a little early to check departure schedules to Alpha Centauri. The neutrino problem has been around for barely a decade; historically, many major scientific controversies have gone unresolved for centuries. Still, it would be nice to have an answer soon, and attempts are being made to find one.

During the course of my research for this book, I talked with solar scientists from one end of the United States to the other. In the summer of 1979 I began hearing what were described to me as "scientific rumors." The thrust of these rumors was that a group of European scientists had solved the neutrino problem. They had supposedly remeasured the nuclear reaction cross sections and found—*mirabile dictu*—that the numbers came out all right after all. The absent-minded scientists were looking for a pair of glasses that had been in their pocket all along. The neutrinos were no longer missing.

Trying to pin down the specifics of the rumors, I consulted neutrino experimenter Bahcall at Princeton. When I repeated the assertion that the neutrino problem had been solved, Bahcall's response was succinct and emphatic: "Nonsense!"

Bahcall went on to explain that a group in Germany had, indeed, remeasured one branch of the nuclear reaction cross sections. Their result implied a value about 1.8 SNUs lower than previous measurements done at Caltech. Even if one accepts the new figure, "that doesn't make too much difference," according to Bahcall. "There is still a very serious problem."

Given the inconclusiveness—or irrelevance—of the German calculations, I asked Bahcall what the current thinking on the problem was. "My current thinking," he replied, with true scientific caution, "is that one has to do another experiment."

BACK TO HOMESTAKE

Another experiment *is* being done. Bahcall, Davis, and an international consortium of more than a dozen scientists are going back to Homestake to try again. This time, they will not be subjected, as they were originally, to fervent sales pitches by coathanger and garment-bag salesmen, who assumed that anyone who purchased 100,000 gallons of cleaning fluid must have a whopping big dry-cleaning plant. The cleaning fluid is no longer necessary, because the scientists are in search of even more elusive game than the high-energy boron-decay neutrinos.

The boron neutrinos were model-dependent. Since all solar models are now under a shadow, the new experiment will attempt to get down to the basics. Regardless of how one models the core, low-energy neutrinos should exist if the basic source of solar power is nuclear fusion. One way or another, the search for low-energy neutrinos ought to be conclusive. "If these low-energy solar neutrinos are detected in a future experiment," write Bahcall and Davis "we will know that the present crisis is caused by a lack of astronomical understanding. If the low-energy neutrinos are shown not to reach the earth, then even many physicists would be inclined to suspect their physics."

The new experiment will use gallium as the target, rather than chlorine. When gallium-71 captures a low-energy neutrino, it transmutes into germanium-71, an isotope with a half-life of 11 days. The neutrino-produced germanium will be measured, as was argon in the original experiment, to determine the number of low-energy neutrinos reaching earth.

The major problem with the new experiment is that gallium is a lot more expensive than cleaning fluid. Gallium is a rare metal that is used in making computer components and, ironically, photovoltaic wafers to transform sunlight into electricity. Gallium currently sells for about $500,000 per ton. The Homestake experimenters need 50 tons of it.

As of this writing, they already have about a ton and a half of gallium, enough to run a small-scale pilot experiment. West Germany is providing 25 percent of the funding, and the United States Department of Energy has chipped in with a million dollars, but much more is needed. Considering the current Congressional attitude toward basic research (some members, notably Senator William Proxmire, seem to think it is a waste of taxpayers' money), the megabucks needed to run a full-scale gallium experiment may be as hard to find as the neutrinos themselves. Nevertheless, scientists around the world are anxiously awaiting the results, as the strange case of the missing neutrinos approaches what may be its final act.

The gallium neutrinos, says Bahcall, "are guaranteed to be there." If they aren't, then "something absolutely fundamental has happened." Whether it has happened to astronomy, physics, or the Sun itself remains to be seen. One doesn't quite know where to put one's money, but it would be nice if the Sun does not turn out to be the guilty party. The Sun already has enough problems.

It's shrinking.

Five: THE INCREDIBLE SHRINKING SUN ... AND COMING ATTRACTIONS

If you listen closely, you may hear an ironic, cosmic chuckle emanating from the general direction of the grave of Hermann Ludwig Ferdinand von Helmholtz. Eighty years after his death, Helmholtz's time may finally have come—time for a last I-told-you-so-but-you-wouldn't-listen laugh. Helmholtz was the nineteenth-century German scientist who theorized that the Sun's energy was produced by gravitational contraction. (See Chapter 1.) After a brief period of acceptance, his theory was shot down in a Bonnie and Clyde barrage from geologists, biologists, and nuclear physicists. Helmholtz was wrong; the Sun was not shrinking.

But now John A. Eddy says that it *is* shrinking.

Helmholtz would probably be amused to learn that not everyone believes Eddy, either. However, he might be somewhat chagrined by the fact that Eddy's discovery is not just another future-shock, high-tech, give-me-a-million-and-I'll-do-some-science breakthrough. Rather, Eddy's theory rests on observations made at the Greenwich Royal Observatory beginning in 1750, seventy-one years before Helmholtz was born. All through Helmholtz's life, as he pondered and theorized, the Sun was unceremoniously shrinking by 5 feet every hour, and the astronomers at Greenwich were watching it as it did so. But it was not until 1979 that anyone noticed.

Even if Helmholtz had been aware of the Greenwich measurements, his theory probably would not have survived. Eddy's shrinking Sun is considerably different than the one Helmholtz hypothesized a century ago. Still, it is ironic that in the midst of today's neutrino-induced confusion, a new theory should resonate with echoes from a simpler past.

Jack Eddy is very familiar with the science of those bygone days. He has worked extensively with data recorded by scientists dating back to the time of Galileo and even earlier. His discovery of the shrinking of the Sun is something of a spin-off from his exhaustive study of the Maunder Minimum, the "spotless Sun era" of the seventeenth century.

Soft-spoken and articulate, Eddy is recognized as one of today's leading solar astronomers. Throughout my research, other scientists frequently referred me to his work. Although he normally hangs his hat at the High Altitude Observatory of the National Center for Atmospheric Research (NCAR) in Boulder, Colorado, I caught up with him at the Harvard/Smithsonian Center for Astrophysics in Cambridge, Massachusetts, where he was spending the year as a visiting professor. Like every other scientist I have interviewed, Eddy had an office about the size of a large telephone booth, overflowing with books, papers, charts, and graphs; it is possible that scientists are attracted to the exploration of space because they have so little of it to work in.

I spoke with Eddy not long after he had announced the discovery of the shrinking Sun at a meeting of the American Astronomical Society in the summer of 1979. His findings had been greeted with a certain amount of skepticism, so I began by asking

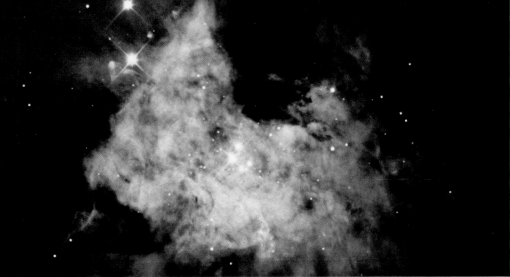

Lick Observatory Photograph

Above: The Great Nebula in Orion—barely visible to the naked eye in the "sword" of Orion—is a spawning ground for stars. This huge cloud of dust and gas is thought to be similar to the solar nebula in which the Sun formed, 4.6 billion years ago. The stars seen here are very young and very hot. Below: The grandeur of a solar prominence is revealed in this photo. Hot gases from the photosphere and chromosphere rise and become trapped in the intense magnetic field lines that dominate the solar atmosphere above the chromosphere.

Naval Research Laboratory

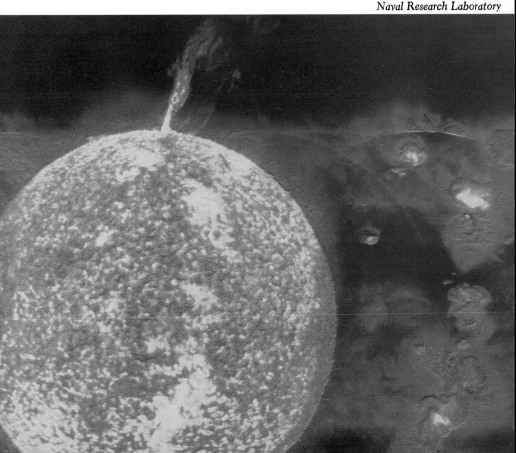

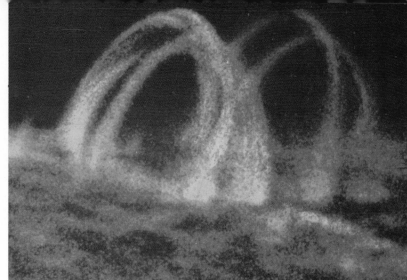

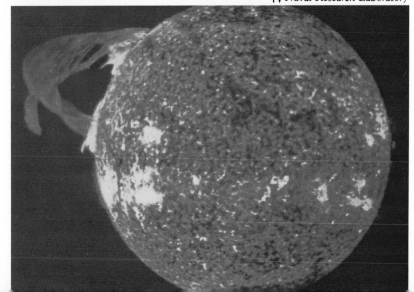

Top left: Magnetic loops are formed by hot ionized gases above active regions. They connect regions of opposite magnetic polarity. Middle left: This "elbow prominence" was photographed from Skylab in UV light. In a temperature range between 30,000 K and 90,000 K, this prominence, barely 20 minutes old, rises more than 600,000 km above the photosphere. The Earth is about the size of the black dot near the base of the prominence.
Bottom left: Another gigantic prominence. This one is about an hour old.

Right: In false-color UV images from Skylab chromospheric temperatures are red, the transition zone is green, and the corona is blue. In just ten minutes, the gases break free from the magnetic loops and become superheated in the corona.

Harvard College Observatory

Harvard College Observatory

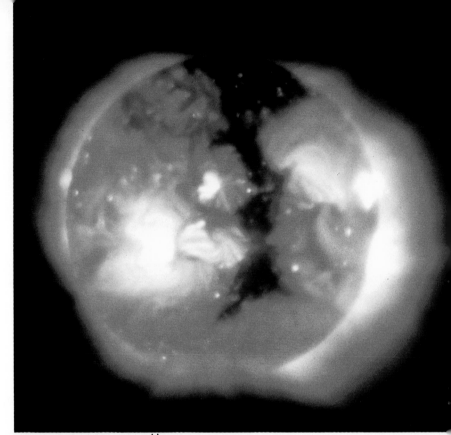

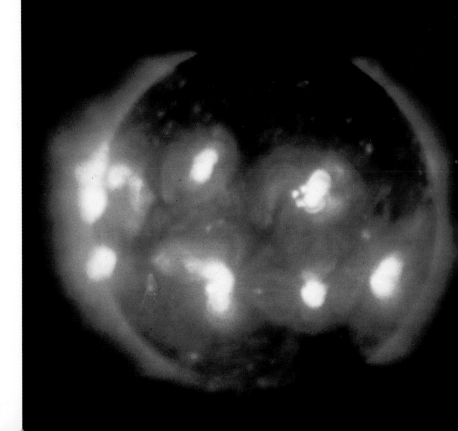

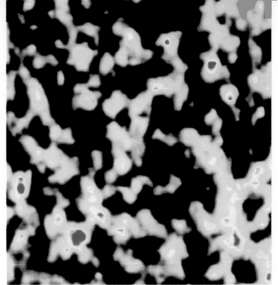

Top left: A coronal hole is revealed in a Skylab X-ray image. Bright active regions appear in sharp contrast to the dark gap in the corona, through which high-speed subatomic particles escape into space.

Bottom left: Active regions in the corona stand out like explosions in this X-ray image. The temperature in these regions exceeds 5,000,000 K.

Right: Seen simultaneously in different UV wavelengths from Skylab, the chromospheric network spreads and blurs with altitude. Low in the chromosphere at a temperature of 20,000 K, the network is sharp and distinct. In the transition region, at 300,000 K, gas pressure declines and the clumps of spicules begin to blur. At coronal temperatures of 1.4 million K, the network all but disappears.

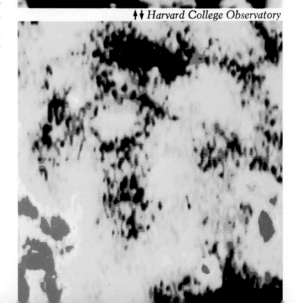

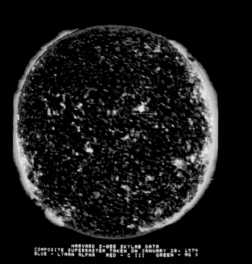

Harvard College Observatory

NASA

Above left: The quiet Sun is seen in a composite of 44 Skylab UV pictures. The hot UV bright points are scattered over the disk like diamonds. The greenish ring around the Sun is the lower corona. Above right: Future space travelers may roam the solar system in immense solar sailing vehicles. Propelled by the pressure of sunlight, the sailers will utilize huge sheets of ultra-thin Mylar to catch the light. A solar sailer in earth orbit would be easily visible from the ground. Below: An artist's depiction of the launch of a Solar Electric Propulsion spacecraft. Carried aloft on the Space Shuttle, the SEP vehicle would fire a chemical rocket to escape Earth orbit, then unfurl large solar panels to collect energy to be used in ejecting mercury ions for propulsion. Thrust would be low but constant, making the spacecraft highly maneuverable.

NASA

The changing face of Jupiter is seen in these Voyager photographs. Taken four months apart by Voyagers 1 and 2, the pictures show the varying rotational speeds of the Jovian cloud zones. The Great Red Spot is a long-lived feature which may be analogous to a hurricane—a hurricane several times the size of the Earth. The Great Red Spot varies in size and shape, and there is some evidence linking these variations with the eleven-year solar cycle. At the lower left in the Voyager 1 image, the giant moon Ganymede is visible.

Voyager—NASA/JPL

Voyager—NASA/JPL

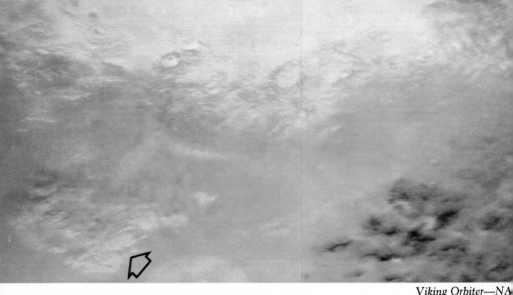

Viking Orbiter—NA

Above: Weather on Mars mainly consists of huge global dust storms. The dust clouds are apparently created by solar heating once each Martian year, during that planet's closest approach to the Sun. This local dust storm (arrow), some 300 km across, is typical. Below: The weather on Venus is consistently hot and cloudy. The dense clouds of CO_2 trap solar radiation, resulting in a global greenhouse effect. Venus rotates very slowly and has no oceans, so the atmospheric circulation is almost entirely solar-driven. Wind speeds on Venus may reach more than 220 miles per hour. Surface temperatures exceed 900°F.

Pioneer-Venus—NASA/Ames Research (

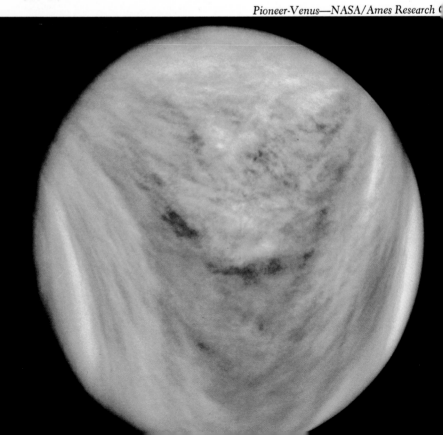

him, was the Sun shrinking, yes or no. "How do I say that with the proper amount of hedging?" Eddy wondered. "I think it *is* . . . at least over the last one hundred years."

Considering the difficulty in obtaining good data on the subject, almost any amount of hedging would be proper. Eddy's work on the Maunder Minimum turned up evidence that the rate of solar rotation was variable. It had been known for more than a century that the photosphere rotates differentially with respect to latitude, with the equator spinning faster than the poles. The differential rotation is the suspected cause of the sunspot cycle, but it had been assumed that the rotation rates were constant.

"When I found evidence that the Sun's rotation rate had changed," Eddy explained, "I wondered if that might show up in its angular diameter. I was also convinced that the luminosity of the Sun might vary slightly with the overall level of solar activity, and I was wondering if we might see that in the luminosity-radius relationship. What I expected to find was that over hundred-year periods the Sun's diameter might rise and fall slightly." To his surprise, Eddy found that the long-term trend was in one direction only—the Sun was shrinking.

Another surprise was the rate of shrinkage. Helmholtz's Sun contracted very slowly, although not slowly enough to leave room for life to evolve on Earth. But Eddy's Sun is shrinking at a rate of 120 feet per day, or 830 miles per century. If it continued to shrink at that rate, in another 100,000 years it would match the dimensions of the 12-inch fiery bellows proposed by Heraclitus 2,500 years ago. That seems unlikely to happen. Sooner or later, the contraction will reverse itself; in the meantime, the evidence indicates that the Sun is continuing to shrink.

That evidence is principally based on observations taken at the Greenwich Royal Observatory and somewhat later observations from the U.S. Naval Observatory in Washington, D.C. The Greenwich observers were interested in establishing precise figures for the dimensions of the Earth's orbit. To that end, they made almost daily measurements with a simple transit telescope, recording the time it took for the full solar disk to pass behind the cross hairs of the instrument. The elapsed time figures could be used to calculate the Earth's orbit, but they also served as an indicator of the angular diameter of the Sun. If the Sun were shrinking, then over the course of decades it would tend to pass the cross hairs slightly faster. That, according to Eddy, is precisely what the old records reveal.

But how reliable are those old records? Can we trust their accuracy? "That's a very good question," Eddy readily admits. "It's one we spent a long time trying to answer, and I think the answer is yes." The instrument used to make the observations was simple and reliable, and Eddy rules out the possibility that the telescope was in some way defective. Similarly, he has confidence in the professional skills of the observers themselves. "Just because they lived a hundred years ago," he says, "doesn't mean they weren't trustworthy."

What does worry Eddy, however, is the quality of the sky through which the observations were made. Greenwich is not exactly a remote observatory in the Andes. During the past century, industrial pollution has filled the atmosphere with tiny particles which scatter sunlight and make the sky seem brighter. "The greater the contrast between two things you look at," Eddy explains, "the bigger the small one looks. That's the reason why fat girls don't wear white dresses. A brighter Sun against a darker sky will appear slightly larger; this is a well-documented effect. So if the sky were gradually getting grungier all these years, as it certainly did at Greenwich, the effect would be to make the contrast go down and the Sun appear slightly smaller. And that's what we see."

Fortunately, there is also data available from the U.S. Naval Observatory, dating back to 1846. The Washington records show roughly the same rate of solar shrinkage as the Greenwich observations. "Therefore it can't just be the dirtying sky over London because the same thing has been happening at Washington, and those are two very different cities, far apart. It seems unlikely," says Eddy, "that they would [become progressively more polluted] in the same way. So unlikely that I think it's just as likely that the Sun is shrinking."

Given the mutually supporting evidence from Greenwich and Washington, Eddy is not reluctant to invoke more isolated data points in support of his theory. The Greenwich records used by Eddy and his co-worker, Boston mathematician Aram A. Boornazian, date back only as far as 1836, but there is at least some evidence that the Sun has been shrinking for more than 400 years.

In 1567, an eclipse of the Sun was visible from Rome. If the

Sun was the same size then that it is today, the eclipse should have been total, but a contemporary observer noted that the moon "did not obscure the whole Sun . . . but a certain narrow circle was left on the Sun, surrounding the whole of the moon on all sides." This account would seem to support the idea that the Sun was slightly larger in 1567.

"I would like to lay my hands on a few more eclipse accounts," Eddy says. "There was a lot of debate in the 1600s about whether the moon was big enough to cover up the Sun at the time of an eclipse. That seemed to me a curious case, and it may be that at the time the Sun was enough larger. But," he cautions, "that's a weak line of evidence."

Some astronomers are not sure how strong the rest of Eddy's evidence is, either. Although Eddy downplays the amount of controversy his theory has aroused, it is fair to say that not everyone is convinced that the Sun really is shrinking. "Basically," said one astronomer, "I'm very skeptical at this point." Another scientist considered Eddy's theory "extremely unlikely."

But Eddy is not alone in his belief. A group of scientists at NASA's Goddard Space Flight Center have studied records of solar measurements made between 1850 and 1937 and come to the conclusion that the Sun *is* shrinking, but at a much slower rate than Eddy's results indicate. The discrepancy seems to lie in the fact that Eddy used horizontal measurements and the Goddard group used vertical measurements. The horizontal diameter measurements were made by clock (timing the transit of the Sun past the cross hairs), while the vertical dimensions were measured with a micrometer, which the Goddard group believes was more accurate.

Taken together, the two sets of measurements would indicate that the equatorial diameter of the Sun is shrinking faster than the polar diameter. Thus, the Sun would be becoming prolate, rather than oblate, as in Dicke's scalar-tensor theory. "But I don't quite believe that," Eddy says. "The absolute value measurements are subject to quite a few errors."

If the Sun truly is shrinking, according to Eddy, "it almost has to be a surface effect." The Greenwich measurements were of the white light photosphere and, of course, revealed nothing about the solar interior. "Anything beyond that is pure speculation. The reasons I say it's just the outer atmosphere and that it has to be oscillatory are based on the rates that are involved. The shrinkage rate is so fast that it would produce an embarrassing amount of light and heat just by gravitational contraction, if the whole Sun's mass were shrinking. Therefore, it's either wrong, or it's not the whole Sun that's shrinking."

To Eddy, there is nothing particularly radical about the concept of a shrinking photosphere. Indeed, based on our knowledge of the behavior of the outer solar atmosphere, the idea makes a great deal of sense. "We know, for instance," Eddy points out, "that the chromosphere bounces all over the place, and the corona is constantly changing. So why is it we think that the photosphere ought to be like a leather ball that doesn't move?"

It seems highly unlikely that anything interior to the photosphere could be involved in the shrinking. If the convective zone and the core were also shrinking at the observed rate, the total solar luminosity would be 200 times as great as it is. The upper limit, then, on the amount of material involved in the contraction is 1/200, or 0.5 percent of the total solar mass. "I suspect," Eddy adds, "that it's a good deal less than that."

Even if the amount is significantly less than 0.5 percent, the fact that any energy at all is supplied by gravitational contraction is important. Energy released by shrinkage is energy which does not have to be accounted for by nuclear reactions in the core. In light of the neutrino shortage, any mechanism which allows a cooler core must be given careful consideration.

"One has to remember, though," Eddy cautions, "that people are very anxious to explain this awful embarrassment about the neutrinos. And I'm guilty of that, too. When I first saw the solar diameter [shrinking] . . . the reason, I must confess, that I got excited about seeing this downward trend was, 'Aha! This'll get us out of the neutrino problem!' But one has to be careful about finding popular solutions."

Eddy does believe, however, that a shrinking Sun helps the neutrino problem. "One way around it," he says, "is to say that the neutrino flux is correct, that the interior temperature is not fifteen million degrees. All we need to do then is drop it down to about fourteen million. That's not so bad; who cares whether it's fourteen or fifteen? But if the interior temperature were fourteen million, we shouldn't be getting as much light from the Sun as we are." The energy provided by shrinkage, Eddy believes, would make up the difference. "The Sun may be getting most of its energy—and the problem is I can't tell you what fraction, but let us say, in round numbers, two-thirds of its energy—from nuclear reactions in a fourteen-million-degree interior. And that is being supplemented on the outside, in the present era, by the contraction of the Sun, which adds another third."

Neutrino-hunter Bahcall has discussed the matter with Eddy but has come to a somewhat different conclusion about the contraction/neutrino connection. "I don't believe it," says Bahcall. "I don't think it has anything to do with the solar neutrino problem." Bahcall's main objection is that if the Sun is shrinking, it must be doing so on a time scale far too short to have any effect on neutrino production.

Eddy recognizes the difficulty. If the Sun is getting a third of its energy from contraction, then at some point in the past, it could only have been two-thirds as bright as it is now. That would return us to the problem of a frozen Earth. The various constraints on solar luminosity suggest to Eddy that the flow of energy from the interior must also be fluctuating, in counterpoint to the contraction and swelling of the outer layers. "What I'm proposing," Eddy explains, "is that as it [core energy production] goes up and down in a very long, slow cycle, the Sun's outer atmosphere shrinks or expands in order to keep the whole thing roughly constant. That is a kind of pious hope, I guess, and may be just as naive as saying that the Sun shouldn't change at all. But it's a possibility."

"What cannot happen," Eddy continues, "is that if the Sun starts to shrink, the signal gets to the middle and therefore the fire goes down. . . . I think it has to be the other way; if the energy from the interior is going up and down, the outer layer adjusts. . . . But that's really going out on a limb at the present, until I can get a better feel for how far back in the past this might go."

Again, Eddy's hedging is entirely justified. He presented his paper on the shrinking Sun at the AAS meeting "because we thought it was kind of exciting," even though he didn't believe he was really ready to publish. "I'm kind of worried now," he admits, "about having to go further in explaining it than I was ready to do. I believe it's shrinking. But I don't have in my hands the rockbound case that I wish I had."

To get his rockbound case, Eddy would like to have data from other observatories besides Greenwich and Washington. There is some data from Russia and South Africa, but the analysis remains to be done. Adding to the frustration of sifting historical records is the fact that although there are records of solar measurements made between 1836 and 1953, the last twenty years are missing. The original Greenwich Observatory shut down and moved to another station, ending the single telescope measurements. The kind of records taken at Greenwich, moreover, are unlikely to be made today because there are more accurate means of measuring the Earth's orbit, which was the original purpose of the observations.

SOLAR OBSERVATIONS: KITT PEAK

Modern solar observatories are less concerned with measuring diameters than with analysis of solar composition, magnetic fields, and rotational velocities. Kitt Peak National Observatory in Arizona is probably the leading solar observatory in the world today; those who work there will tell you that it is the finest observatory of *any* kind. The work carried on at Kitt Peak is a far cry from the almost quaint observational programs conducted at Greenwich a century ago.

Kitt Peak is open to the public, and, to date, over a million people have toured the facilities. Located in the Quinlan mountains about forty miles west of Tucson, Kitt Peak was selected as the site for a national observatory after an exhaustive study in the 1950s. The mountain itself reaches some 6,500 feet above sea level and towers nearly a mile above the surrounding scrublands. The land is owned by the Papago Indians, who lease the peak to a consortium of fourteen universities. The Indians were originally reluctant to let the astronomers use the site, since the mountain is the sacred home of one of their gods, Eel-ol-top. But after being treated to a look at the moon through a University of Arizona telescope, the tribal council was happy to grant the lease to "the men with the long eyes." It seemed that Eel-ol-top's primary responsibility was to watch over the heavens, a task completely compatible with the astronomers' goals.

Today, there are fifteen major telescopes atop Kitt Peak. The largest is the 158-inch Mayall Telescope, housed in a nineteen-story white dome that is visible from the desert 30 miles away.

But the main attraction is the McMath solar telescope, a unique structure that vaguely resembles the set of a James Bond movie. The McMath Telescope is actually built into the mountain; half of its 500-foot length is underground. The visible portion angles upward to a 100-foot support tower.

The McMath Telescope is devoted mainly to the observation of the Sun. An 80-inch mirror mounted at the top of the shaft tracks the Sun and reflects the light all the way to a 60-inch mirror at the bottom of the shaft. From there, the image is directed to a 48-inch mirror, which reflects it downward into the underground observation room. There, a 34-inch image of the Sun is displayed on a massive viewing table. A wide variety of filters and spectrographic equipment may be employed to emphasize particular aspects of the Sun. The underground viewing room looks totally unrelated to the study of astronomy, in fact, more like the operations room of an aircraft carrier. The days of the solitary man of science peering hopefully into an eyepiece while the rest of the world sleeps are apparently gone forever.

Current work at Kitt Peak is devoted to such problems as solar magnetism, rotation rates, and oscillations. Another continuing program is the determination of the isotopic composition of the Sun. Early investigators had little trouble in identifying the distinctive spectrographic signatures of elements such as calcium, sodium, and hydrogen, but, as the coronium confusion demonstrated, other elements may be more subtle. Theoretically, it should be possible to identify all the more than 100 elements in the solar spectrum, since all of them should be present in the Sun. Since different isotopes of the elements decay at different rates, analysis of the solar composition gives important clues to the makeup of the original solar nebula. The presence or absence of particular isotopes may be crucial indications of the kinds of processes at work in the Sun. Lithium, for example, is an element of particular interest. Its spectral lines are extremely weak, but if lithium is found in the spectrum, that would be strong evidence that material at the surface of the Sun does not circulate to the core. The absence of lithium would suggest that convection does carry surface material to the core, where the lithium would quickly burn up.

With such important work being carried on, it is difficult to get viewing time on the McMath Telescope to investigate new theories, such as solar shrinkage. Typically, a visiting astronomer must apply for telescope time more than a year in advance. In addition, any viewing program designed to check for shrinkage would probably have to be conducted over the course of many years. That, in turn, would undoubtedly necessitate studies of the long-term quality of the atmosphere over southern Arizona.

SOLAR OBSERVATIONS FROM SPACE

The ideal place to measure the Sun's diameter is from space, where there are no problems with the atmosphere. Although the Skylab astronauts took over 150,000 high-quality pictures of the Sun, unfortunately none of them was applicable to the study of such important problems as Hill's oscillations or Eddy's shrinking. The Skylab telescope took a number of white-light images of the photosphere, but, according to Eddy, "it wasn't taking them frequently enough to measure what Henry [Hill] does, nor accurately enough to measure what we do."

Even though past observations from space offer little help, Eddy's eyes light up when he contemplates the future. Over the next decade, NASA hopes to launch at least two missions which will almost certainly revolutionize solar science. One, the Solar Polar mission, will view the Sun from above its north and south poles; the second, known as the Solar Plunger, will actually penetrate the outer corona.

Solar Polar is a two-spacecraft mission, now scheduled for launch aboard the Space Shuttle in early 1983.* All previous probes have viewed the solar system from near the plane of the Sun's equator. Because the Earth is not a stationary platform, any vehicle launched from it carries with it the Earth's orbital velocity of some 66,000 miles per hour. To get out of the orbital plane, a spacecraft would have to expend more energy overcoming its Earth-imparted velocity than is possible with contemporary boosters. But NASA planners have designed an ingenious method of flinging the Solar Polar spacecraft out of the equatorial plane. They intend to use the planet Jupiter as a gigantic slingshot.

* Solar Polar, like all proposed missions, is at the mercy of Congress. Budget-cutting has been particularly devastating for NASA's unmanned exploration program. In addition, technical problems in the Space Shuttle have caused delays. Solar Polar, if it gets off the ground at all, probably won't be launched until the mid-Eighties.

The gravity assist technique has been used before, most notably on the Voyager missions to the outer planets. As a spacecraft approaches a planet, gravitational forces cause it to accelerate. Following a carefully plotted trajectory, the spacecraft picks up enough speed so that it can whip past the planet and sail away in a completely different direction. The Voyagers used Jupiter's gravity to swing them on toward Saturn. At Saturn, Voyager 2 will repeat the technique and continue on to Uranus; if the spacecraft is still healthy at Uranus in 1986, it may use the gravity assist procedure one more time for a trip to Neptune, arriving there in 1989.

Even more exotic gravity assist missions are under consideration. A proposal for a 1989 Saturn mission suggests a bizarre double gravity assist from the Earth itself. The spacecraft will be launched into an elliptical orbit which will take it outward as far as Mars before it falls back toward Earth for its first gravity boost. Its next loop will take it farther out, and then back in for a second gravity assist which will impart enough energy for the spacecraft to reach Saturn in 1997.

Solar Polar will send two vehicles to Jupiter for a gravity assist. One will be flung around Jupiter and then upward, north of the solar equator; the other will take a southern route. Both spacecraft will then loop back toward the Sun and observe it from a vantage point never before possible.

The spacecraft will carry more than a dozen sophisticated instruments to study not only the Sun, but the environment outside the equatorial plane. Of special interest are the magnetic field lines from the Sun, which dominate the interplanetary medium. Because of the long-lived coronal holes over the Sun's poles, observations of the nonequatorial solar wind should also be valuable. The spacecraft will also study x-ray and gamma-ray bursts from the Sun, radio emissions, charged particles, and galactic cosmic rays. All things considered, Solar Polar should be a veritable scientific smorgasbord, and solar experts are already salivating in anticipation.

The Solar Plunger mission has yet to receive funding, but NASA and the Jet Propulsion Laboratory (JPL) in Pasadena, California, have put together a massive 564-page study of the mission's requirements. The study includes papers with such titles as: "Measurement of Solar Gravitational Oblateness with Gravity Gradiometers," "The Determination of the Structure and Heating Mechanisms of Coronal Loops from Soft X-Ray Observations with the Solar Probe," "Solar Neutron Spectroscopy Near the Sun," and "Detection of Gravitational Radiation and Oscillations of the Sun via Doppler Tracking of Spacecraft." Clearly, the Solar Plunger is being designed to answer some very fundamental questions about the Sun.

If approved, the Solar Plunger would be another Earth double gravity assist mission. A 1986 launch would make two large Earth orbits, then proceed to Jupiter for a final fling back into the Sun. The spacecraft would pass within 4 solar radii—or less than 2 million miles—of the sun in 1991. Such a trajectory would take it into the outer corona.

Obviously, such a close encounter with the Sun will involve special problems. The entire spacecraft will have to be protected by an umbrella-like heat shield. Instruments will make observations through diamond windows. The communications system will have to be guarded against disruption by solar activity.

But these are merely technical problems. The biggest problem is simply getting the money for the mission. Throughout the 1960s and 1970s, NASA launched swarms of unmanned probes—Mariners, Pioneers, Vikings, Voyagers—to explore the solar system. But by the mid-1970s, Congress was less receptive to the attractions of the cosmos, and one mission after another was trimmed from the NASA budget. The result is that there will be a three- or four-year hiatus in the mid-1980s, with no new missions.

At the 1979 meeting of the American Astronomical Society's Division of Planetary Science, the mood was gloomy and desperate. Scientists who had spent their entire careers investigating the solar system were now facing the prospect of virtual unemployment. NASA officials tried to soothe their fears with optimistic projections of a space renaissance in the late 1980s, with missions to comets, asteroids, and the moons of Saturn, but not one of these proposals has yet received funding.

As resources become scarce, competition increases. Geologists lobby for Mars Rovers; experts in particles and fields urge more missions to Jupiter and Saturn; atmospheric specialists long for Venus probes; each has reasons why his pet mission is of fundamental importance. The infighting is restrained but deeply felt. Thus, some planetary scientists have been heard to denigrate the Solar Polar, calling it "a mission to nowhere." Inevitably, selling one mission means minimizing the importance of others. The net result is that missions such as the Solar Plunger may never get off the ground.

But help may be on the way. The era of the Space Shuttle is about to dawn, and if the system works as well as advertised, the cost of space exploration will be reduced. Further, the United States no longer has to pick up the entire tab. The European Space Agency, a consortium of eleven different nations, is playing a larger role in the design and building of spacecraft. Japan and Canada are also making contributions, and, of course, the Soviet Union has the capacity, if not always the desire, to participate in space exploration. And some space enthusiasts are hopeful that the oil-rich Arab nations may want to invest some of their booty in the study of the Universe.

SOLAR PROPULSION

If the Solar Plunger does become a reality, it will not only study the Sun, it will use it. The power source for the spacecraft will be Solar Electric Propulsion (SEP), also known as ion drive. An array of solar collectors will concentrate the energy of sunlight and use it to accelerate mercury ions to velocities as high as 45 kilometers per second. The ejected ions, via Newton's third law, will shove the spacecraft forward.

There are advantages and disadvantages in using SEP. The major disadvantage is the extremely low thrust it provides. SEP missions will take much longer to get to where they are going; the complicated double and triple gravity assist trajectories will be necessities with SEP. On the other hand, although the thrust is low, it is constant. Rather than expending all their fuel in one or two big burns, as in chemical rockets, SEP vehicles will have a continuous power supply. As a result, they will be highly maneuverable. NASA planners envision a SEP mission to the asteroid belt, in which a spacecraft will wander among the asteroids like a beachcomber with a metal detector, changing trajectories with ease.

Ion drive was chosen as the propulsion system of the future after an intramural competition with an alternative solar-powered vehicle. During the mid-1970s, scientists and engineers at the Jet Propulsion Laboratory did a fascinating study of so-called Purple Pigeon Projects—a variety of exotic proposals for medium-to-distant future space exploration. Ion drive's principal rival was solar sailing.

The idea of solar sailing is almost irresistible. A solar sailing vehicle would consist of a spacecraft attached to an immense Mylar sail, kilometers wide but only microns thick. Sunlight consists of photon particles, and when photons strike something they exert pressure. They would impart enough force to a solar sail to propel the vehicle through interplanetary space. As with SEP, acceleration would be slow but constant—and free.

Solar sailers would ply the spaceways as clipper ships of old plied the oceans. They would use the sunlight in precisely the same way that the old vessels used the wind. A solar sailer could, for example, fly to Mars, drop off a lander, wait in orbit while the lander collects soil samples, rendezvous with a sample-return probe, and then tack inward, back to the Earth. The sailers would be reusable and cheap. Science writer Jonathan Eberhart has neatly summed up the appeal of solar sailing: "If there is ever a *Whole Space Catalog*," he predicts, "it will be delivered by solar sail."

The JPL study ultimately chose SEP over solar sailing, at least in part because ion drive is simply a variation on existing technology, while solar sailing would be a mind-boggling departure. A Congress reluctant to dole out money for relatively tiny vehicles would almost certainly balk at the idea of constructing mile-wide solar sails in orbit. Yet there are many who believe that in the long run—meaning the twenty-first century—solar sailing may prove to be the most practical and economical method of travel within the solar system. With a little imagination, one can even foresee solar yachting regattas, Earth-Moon races, and leisurely vacation cruises to Mars. People alive today may see it happen.

For the immediate future, however, the romance of sailing the spaceways must take a back seat to the imperatives of science. Every time we have flown a mission to one of the planets, we have been surprised. Our understanding of the solar system has undergone a sweeping revolution in the last twenty years. Solar Polar and Solar Plunger will almost certainly have the same effect on our understanding of the Sun.

Today, solar science is in a state of near-chaos. We aren't sure if the Sun is shrinking. We don't understand how solar magnetic fields interact with flares and transients. We don't fully comprehend the mechanisms involved in the creation of the solar wind. And we are not even sure how the Sun produces its energy. Answering these questions, and many others, may take decades, but the effort will be worthwhile. The Sun may be a small, obscure star, but it is the only one we have. And like the Earth, it does not come with an owner's manual; we have to figure it out for ourselves.

Six: THE INCONSTANT SUN

The more we learn about the Sun, the less constant it seems. We have fragments of evidence for solar variability from a dozen different areas of scientific research, but the fragments have yet to be arranged into a single meaningful picture. How does one weigh and balance the evidence from Martian riverbeds, Antarctic ice fields, coral reefs, oceanic sendiments, tree rings, upper atmospheric chemical reactions, and lunar pebbles? How significant are models of the young Sun of four billion years ago, when we don't even understand what the Sun is doing today?

The question of solar variability is one of the most controversial in the entire volatile field of solar science. There are respected scientists who see in the evidence virtually no indication of changes in the solar output. Equally respected scientists, examining the same body of evidence, see strong hints of significant solar variability. The interested layman following the debate might well conclude that nobody is really very sure of anything.

In an attempt to make sense of it all, one can identify the general areas of study and list some of the specific evidence under consideration. The list that follows is arbitrary and by no means complete. Categorization is ambiguous; are moon rocks astronomical evidence or geological? One could easily reshuffle the evidence and come up with completely different categories. With these caveats in mind, consider the nature of the evidence for solar variability.

Astronomical evidence: Stellar evolution and behavior; galactic rotation; galactic cosmic-ray flux; outer-planet brightening; terrestrial orbital defects; meteorite composition; lunar evolution

Atmospheric evidence: Early atmosphere composition; atmospheric evolution; carbon-14 content; ozone balance; auroral displays; greenhouse effect; atmospheres of Venus and Mars

Biological evidence: Origin of life; evolution of photosynthesis; waves of extinctions; coral reef composition; tree ring growth

Geological evidence: Martian riverbeds and layered icecaps; geomagnetism; composition of lunar rock samples; Antarctic and Greenland glacial cores; sea bed cores

Paleoclimatic evidence: Ice ages; periodic droughts; short-term climatic variations; sun-weather links

Physical and theoretical evidence: Neutrino shortage; solar core modeling; solar dynamo modeling; isotopic composition of solar wind; extinct radionuclides; solar and planetary magnetic fields

Observational evidence: Sunspot cycles; historical records; solar rotation; eclipse observations; spacecraft solar wind measurements; spectral variations; solar oscillations; direct measurements of the solar constant

If the Sun's energy output ever varies significantly, its impact would be registered, one way or another, in most of the fields listed. With so many avenues of analysis, one might be led to believe that it would be relatively easy to find proof of solar variability; in fact, the case is precisely the opposite. Solar variability cannot be pinned down because the evidence itself is variable. By analogy, imagine that the head referee of the Superbowl is found dead on the 50-yard line. In the general confusion that ensues, the body somehow disappears—no autopsy is possible. There are a hundred thousand eyewitnesses, yet no one is sure what actually happened. Some people say they heard a shot fired. Others heard no shot and maintain that the referee must have died of a heart attack. Still others claim they saw a knife sticking out of the unfortunate official's back. Television replays are ambiguous; different camera angles seem to reveal different events.

So what actually happened? Was the referee murdered? More to the point, is solar luminosity variable? The answer is that it depends on what evidence you choose to believe.

Much of the confusion and controversy results from the fact that almost all of the evidence for solar variation is indirect evidence. We have no direct measurements of the solar output a billion years ago—or even a hundred. Measuring the solar constant, S, is a kind of scientific drudgery, akin to counting railroad cars and compiling telephone directories. As Eddy has pointed out: "It is a difficult measurement; it is an unexciting measurement; it requires dedication and funding over periods measured not in years but in decades." Not surprisingly, then, our record of the solar constant is spotty at best, even for the immediate past.

Relying on indirect evidence, we are forced to make inferences about the meaning and significance of that evidence. This is seldom easy or straightforward. If we have evidence of an ancient event, X, we can only speculate about the mechanisms which might have caused X. One can build a case for mechanism A, which involves changes in the solar constant, but there may also be mechanisms B, C, and D, which do not involve the Sun. At this point, human prejudice begins to get in the way. If we are primarily interested in finding changes in the solar constant, then there will be a tendency to concentrate our attention on mechanism A, to the exclusion of the others. The result may be a seemingly convincing case for A $\rightarrow$ X. But nature doesn't share our biases, and the truth may be that B $\rightarrow$ X or C $\rightarrow$ X, or even A + B + C + D + E + F $\rightarrow$ X.

Solar variability is slippery because it may involve dozens of mechanisms and time scales ranging from seconds to billions of years. The short-time-scale variations are difficult to pin down because any such changes would have to be extremely small. In cases where the variation is at the level of a few tenths of 1 percent, it may be all but impossible to separate the data from the "noise." Variations with a period of years or decades may also involve relatively tiny changes which strain the limits of observations. Scientific history is littered with the bones of theories based on observations made at the boundaries of resolution; as resolution improves, many of the old theories fall by the wayside. Thus, theories of solar variation which rely on data taken even a few decades ago tend to be suspect. At the other end of the scale, solar variation on time scales of thousands to billions of years depends solely on indirect evidence which is open to many differing interpretations.

Yet another complication is caused by the fact that the solar constant is produced by the total energy output of the Sun, from one end of the spectrum to the other. A variation in solar UV flux, for example, might have no readily discernible effect on the output of visible light. A measurement of the solar constant is, in reality, not one measurement, but many.

With such a wide variety of factors to consider, there is little wonder that the subject of solar variability is so controversial. Still, there does exist a significant body of evidence which seems to point toward solar variations on both long and short time scales. Some of the most intriguing evidence comes not from the Sun, but from the Moon.

More than a decade after the first lunar landing, most people's memories of the Apollo program seem to consist of hazy images of astronauts bounding around on the gray lunar plain before an unflapping flag, gee-whiz quotes, and Nixonian pronouncements on the meaning of it all. The last astronaut left the moon in 1972, and there is no indication that anyone will be returning to it in this century. All that remains from that era when we used to go to the moon is a somewhat faded glory and a few hundred pounds of rocks.

MOON ROCKS

With little or no fanfare, scientists are continuing to analyze those lunar souvenirs. Moon rocks and meteorites are our only direct links with the rest of the universe, and the meteorites have all been contaminated, to some degree, by contact with the Earth.

Astronauts Edwin Aldrin and Neil Armstrong set up a number of scientific experiments during their historic landing on the moon on July 20, 1969. Here Aldrin is standing near equipment designed to study solar wind composition. The Lunar Module is behind him.

The moon rocks, however, are pristine samples of the cosmos, and they tell a fascinating tale.

A rock sitting on the surface of the moon is exposed to direct bombardment by particles from the Sun. Typically, a lunar rock spends about 5 million years on the surface before being reburied by some lunar "gardening" mechanism such as meteorite impact. During its time on the surface, a rock gets blasted by particles from the solar wind, solar flares, and galactic cosmic rays. These impacts leave microscopic craters, colloquially known as "zap pits," which scientists can "read" in much the same way that geologists read terrestrial fossils.

Reading the lunar record, however, is a complicated business. The history of a particular rock sample can only be inferred. No one knows precisely when a rock was exposed, or for how long. Samples from different landing sites may have very different histories, making comparisons difficult. It is not surprising, then, that the lunar record is ambiguous.

Particles from the solar wind and solar flares may consist of the nuclei of elements being produced in the Sun. Upon collision, some of the nuclei may be implanted in a moon rock, leaving a measurable record of the rates of solar particle production. The depth of implantation gives an indication of the energy level of the incoming solar particle.

Today, the average speed of the solar wind is about 400 kilometers per second. Particles impacting on moon rocks at that velocity should (and do) leave characteristic signatures. But a group of French scientists analyzing Apollo samples have found indications that at various periods during the past, some types of solar wind particles have had velocities of only 250 kilometers per second. They conclude that the contemporary solar wind is atypical.

Other scientists, looking at different aspects of the solar wind, agree that the current solar wind intensity is atypical, but in the other direction. R. N. Clayton of the University of Chicago has analyzed the abundance of nitrogen isotopes in lunar samples and believes that some 2 billion to 4 billion years ago the solar wind was about five times as strong as it is today. Scientists at Rice University have come to a similar conclusion, based principally on the lunar abundance of argon isotopes. On the other hand, studies of the rare gas xenon have led some scientists to conclude that the ancient solar wind was stronger than today's, while others see little evidence for xenon-based variations in the early solar wind.

Solar flare particles also produce lunar impacts, but of a higher intensity than those of the solar wind. The typical distance of penetration by a solar wind particle in a moon rock is only about five millionths of a centimeter; flare-accelerated particles penetrate to depths ranging from millimeters to a few centimeters. As with the solar wind, there is considerable disagreement about whether or not the flare impacts point toward solar variations.

J. N. Goswami and a group of researchers from the Physical

Research Laboratory in Ahmadabad, India, have done extensive analyses of the tracks left in lunar rocks by solar flare protons and heavy-element nuclei. Their conclusions are mixed. They believe that during the last several million years, the solar flare flux may have been as much as ten times as small as the contemporary value. For longer periods, however, ranging several billion years into the past, the Indian group believes that the flare rates and intensities could not have been lower than the rates of the more recent past. In addition, they see evidence for a considerably higher level of flare activity in the very early history of the solar system.

E. L. Fireman, of the Smithsonian Astrophysical Observatory, has examined flare-implanted nuclei of argon in lunar rocks. He concludes that the solar flare flux over the last 1,000 to 10,000 years must have been higher than that recorded for the last three solar cycles. His results are supported by the work of H. A. Zook of NASA, who sees at least tentative evidence from the lunar rocks that flare activity between 10,000 and 30,000 years ago was dramatically higher than the present-day norm.

Not all scientists share these conclusions. Ghislaine Crozaz of Washington University believes that while solar flare activity has been continuing for at least 4 billion years, there is no firm evidence that the level of activity was ever much higher than it is today. Crozaz's colleague at Washington U., Ernst Zinner, sees little data to support any change in the flare flux over the last million years. Crozaz, in fact, doubts that it is even possible to be certain of high flare activity in the early history of the solar system. The best samples to examine for such evidence, according to Crozaz, are not the scattered lunar rocks collected by the astronauts, but rather the core samples they excavated from several feet below the lunar surface. But the core samples are exceptionally difficult to "read," and any conclusions based on them would have a high degree of uncertainty.

In addition to solar wind and flare impacts, moon rocks are also subjected to bombardment by galactic cosmic rays. The cosmic rays are particles of extremely high energy, emanating from somewhere out in the depths of the galaxy. Their energy is so high that they penetrate the lunar surface to a depth of meters, rather than millimeters. By mechanisms which are not fully understood, the Sun somehow modulates the rate and intensity of the incoming galactic cosmic rays. It is thought that during periods of high solar activity, solar particles and magnetic fields act as a kind of shield, warding off or decelerating the galactic particles. Theoretically, the record of cosmic-ray tracks in lunar samples ought to give an indication of solar variability. However, the evidence examined to date seems to show no variation in the cosmic-ray flux over the past 50 million years.

It should be apparent that the lunar record has yet to provide us with anything resembling a smoking pistol. There are intriguing hints of significant variations in the solar output, but there is also evidence which points in precisely the opposite direction. One factor which complicates the investigation is our uncertainty about the exact location of the Moon itself. Calculations concerning the lunar zap pits are based on the assumption that the Moon is 1 AU from the Sun. That is certainly true today, but it may not have been true 4 billion years ago. The composition and density of the Moon are such that some cosmologists would feel a lot more comfortable if the Moon had originally formed somewhere inside the present orbit of Mercury. That would make the Moon fit better with the solar nebula theory, but it raises the sticky question of exactly how and when the Moon was captured by the Earth.

The evidence from meteorites is similarly ambiguous. The uncertainties inherent in meteorite analysis are even greater than those associated with studies of moon rocks. Wherever the Moon may have been when it formed, we can be relatively sure that for the last three or four billion years it has been comfortably in orbit around the Earth. But meteorites have spent the years since creation wandering in eccentric orbits. The meteorites are thought to have formed in the orbit of the asteroid belt; however, as Harvard astrophysicist A. G. W. Cameron points out, "The scenario is in a process of constant evolution." It now seems probable that at least some types of meteorites formed in the inner solar system.

Wherever they formed, the interior of meteorites may contain gases from the original solar nebula. Despite problems with terrestrial contamination, scientists have found meteorite samples extremely valuable in establishing some of the basic parameters of the nebula, such as temperature, pressure, and composition. Using meteorites to measure more recent solar activity is less productive. Crozaz asserts that meteorites can probably never give definitive evidence of solar variations; however, they do seem to confirm a general *lack* of variation over the long haul. Predictably, though, there are scientists who disagree with that conclusion. A group at the Johnson Space Center at Houston, Texas, has found meteoric evidence that, between 1967 and 1978, there was substantial solar modulation of galactic cosmic-ray tracks in meteorites. Other workers have analyzed older meteorite samples and have turned up indications that the cosmic-ray flux was much higher during the Maunder Minimum in the seventeenth century. That would be consistent with an overall decline in solar activity during the sunspot hiatus.

THE PLANETS AND THE GALAXY

The solar system, of course, contains much more than just moon rocks and meteorites. Any substantial variation in the solar constant ought to have an effect on the other planets. Observations of Uranus, Neptune, and Saturn's giant moon Titan have provided some intriguing indications of short-term solar variations linked to the 11-year solar cycle. The albedo, or reflected brightness, of these bodies apparently increases and decreases slightly over the course of the 11-year cycle. Titan, closest to the Sun, brightens by about 6 percent, Uranus by 3 percent, and distant Neptune by 2 percent. The differing rates of brightening suggest that these bodies are registering a change in the nonvisible part of the solar spectrum, most probably in the ultraviolet range. All three bodies possess dense atmospheres in which photochemical processes are probably at work. The solar variations may change the rates of the chemical reactions in their atmospheres, thus altering cloud densities and albedos.

The three planets of greatest interest to those searching for variations in the solar output are Venus, Mars, and of course Earth. All three are dense, rocky worlds with evolving atmospheres; presumably, any solar event affecting one would have a similar effect on the others. However, each planet is unique, and direct comparisons are difficult.

As known today, Venus is a literal hell of a planet, with dense clouds of CO_2, rain composed of sulfuric acid, and surface temperatures exceeding 900°C; Mars is dry, frozen, and protected by the thinnest of atmospheres; and Earth, for all its environmental problems, is the Boardwalk of planets, the only prime real estate in the neighborhood. Yet current thinking holds that, four billion years ago, all three planets started out in much the same way. Solar variability may account for some of the changes since then, but so many other factors are involved that even apparent Sun-related correlations are suspect.

Mars, for example, has undergone dramatic changes of climate. Pictures from Mariner 9 and the Viking Orbiters have revealed a planet-wide system of narrow channels which resemble dry riverbeds. At the present time Martian conditions prohibit the existence of much liquid water on the surface, so the implication is that several billion years ago, Mars was much warmer and had a denser atmosphere. A hotter Sun may have been the cause, but other explanations, involving atmospheric evolution, volcanism, and orbital defects, seem more likely.

The Martian polar caps are also of interest. Viking images show distinct layering in the windswept ice fields, indicative of many different episodes of glaciation. If some temporal correla-

This mosaic of Viking Orbiter photos shows the windswept northern polar cap of Mars.

NASA

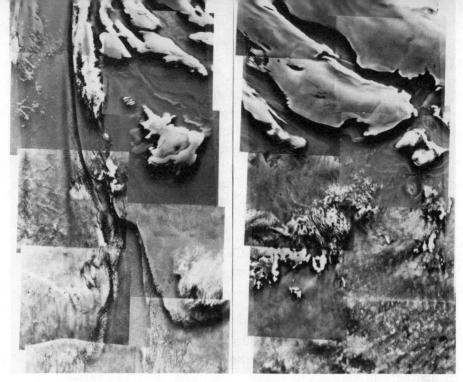

A closer look reveals swirls and terraces indicative of many different episodes of glaciation on Mars. If Martian ice ages coincided with terrestrial ice ages, it would be strong evidence of a solar-climate link.

Viking Orbiter—NASA/JPL

Another face of Mars reveals the planet's watery past. Today, atmospheric pressure is too low to permit the existence of much liquid water on the Martian surface. But these pictures of the Martian region Chryse, near the landing site of Viking 1, clearly show ancient stream beds and indications of catastrophic flooding. Perhaps two to three billion years ago, the Martian climate was warmer and wetter; solar variation is one possible explanation.

Viking Orbiter—NASA/JPL

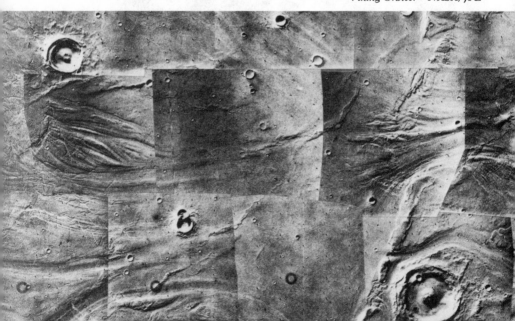

tion could be found that linked the Martian ice ages with Earth's, that would be very strong evidence of solar variation. A decline in the solar output would seem to be the best explanation for concurrent ice ages on different planets.

An alternative explanation does exist, however. The determining factor in planetary temperatures is not the amount of energy released by the Sun, but, rather, the fraction of that energy which reaches the planet. If something got in the way, blocking some of the energy transfer, the cumulative effect would be about the same as if the solar output had been reduced. And periodically, something *does* get in the way.

Just as the planets revolve around the Sun, the Sun itself revolves around the center of the galaxy. A galactic year is equal to about 300 million terrestrial years. Some scientists find it significant that 300 million years also seems to be the average time between major epochs of glaciation on Earth. It is possible that the Sun periodically encounters dense clouds of gas and dust during its journey around the galaxy. If a cloud were dense enough, it might block some of the Sun's energy output, reducing temperatures on Earth and Mars.

This theory has its attractions, but it also has more than its share of complications. Paradoxically, an encounter with a galactic cloud might serve to *increase* the Sun's intrinsic luminosity. The in-falling matter from the cloud would act as a kind of super-charger, temporarily injecting fresh fuel into the solar fire. That would seem to eliminate dust clouds as a cause of glaciation, but yet another paradox is that, by some theories, ice ages may be caused by a temporary rise in solar luminosity. Briefly, an increase in luminosity would melt the polar caps, increasing the percentage of the globe that is covered by water. That, in turn, would increase the amount of solar radiation that is reflected back into space, resulting in a net decline in global mean temperature.

Most theories, however, hold that the reverse is true; a decline in luminosity is needed for glaciation. The sober fact is that nobody is entirely sure about how or why ice ages happen. Galactic dust clouds and solar luminosity may have an effect, but so many other factors are involved that it is probably pointless to talk about ice ages and solar variations without getting into the entire complex realm of Sun-climate connections. We address that subject in Chapter 7; for the moment, our search for solar variations should center on the Sun itself.

THE HISTORICAL RECORD

Direct observations of the Sun can tell us nothing about the possible long-term variations in solar luminosity, and it is even doubtful that we have meaningful data extending more than two or three centuries into the past. Before the invention of the telescope at the beginning of the seventeenth century, naked-eye observations could catalog only a limited number of solar phenomena. Sunspots are sometimes visible to the naked eye, and there is a very incomplete record of them for the last two thousand years. The solar corona can be observed during eclipses, but there is no detailed account of it prior to 1715. The auroras—the northern or southern lights—are atmospheric phenomena related to solar activity, but although they are easily observed, the historical record of them is insubstantial.

Ever since the discovery of the 11-year sunspot cycle, researchers have tried to find in the historical record some evidence of the cycle's existence in antiquity. There are indications of periodic sunspot activity before 1600, but the data are so incomplete that one has to be cautious about drawing conclusions from them. We tend to be biased toward an 11-year cycle because that is the condition which exists today; the discovery of a similar cycle in the old records may simply be a reflection of our own prejudices. Further, the reliability of the sky watchers of yesterday is open to some question. European observers, after all, apparently did not even notice the Crab Supernova of 1054.

With these cautions in mind, we can say that *a* sunspot cycle (not necessarily the present one) has existed for at least the last 2,000 years, and probably for much longer. But it is also clear that the nature of the cycle changes, and there have been some extended periods, such as the Maunder Minimum, when there was little or no sunspot activity. There is some reason to believe that the solar constant has varied with the sunspot excursions, but almost certainly by less than 1 percent.

Auroral displays are caused by the interaction of charged solar particles with the Earth's ionosphere. During periods of high solar activity, auroral displays are more frequent and intense. But again, the historical record before about 1700 fails to give a positive indication of an indisputable solar cycle. We do know that during the Maunder Minimum auroral displays were extremely rare; indeed, their reappearance at the end of the Minimum was a

source of consternation for those who were unaware of the existence of the phenomenon. There are accounts of auroral activity
before the Maunder Minimum, and taken at face value they
would indicate that solar activity was much lower then than it
is today. This may be a real effect, or it may simply be another
indication that, before the Renaissance, nobody was much interested in the sky.

Eclipses are impossible to ignore, and they are well documented
throughout history. We know that the size and shape of the corona varies with solar activity, and it is a spectacular sight which
even the humblest pre-Renaissance serf could not have failed to
notice. Yet the curious fact remains: we have no unambiguous
record of the modern, structured corona prior to 1715.

Eddy's theory of the shrinking Sun could be one explanation
for the lack of corona accounts. A larger Sun might not be totally
eclipsed by the moon, and the corona would consequently appear
much dimmer. However, the mean Earth-Moon distance at the
time of an eclipse varies, affecting the extent and duration of totality; only a very small number of eclipses would fall into the range
where the shrinking Sun factor would be important.

It is also possible that the corona, as we see it today, simply
did not exist before 1715.

Eddy finds it a "possibly significant coincidence" that the corona was first reported at about the same time as an abrupt increase in auroras and sunspots. All of this, according to Eddy,
"could mean that the bright and structured corona is indeed a
modern feature of the Sun which first appeared at the time of the
Sun's recovery from the Maunder Minimum."

Drawing grand conclusions from the historical record is risky,
but we cannot simply ignore what evidence we do have. Eddy
concludes that "during the last 100 or 200 years, when we have
observed the Sun most intensely, its behavior may have been
unusually regular and benign." This is a somewhat troubling conclusion, because it implies that if the Sun has behaved strangely
in the recent past, it may do so again in the immediate future.

THE SOLAR CONSTANT

One reason it is so difficult to deduce changes in the solar constant in the past is that we don't even have very precise measurements of the solar output today. As mentioned earlier, measuring

the solar constant is a dull, tedious, and meticulous exercise and not one which many scientists are inclined to undertake. However, there are a few recent measurements, as well as signs that a nearly continuous solar watch will be instituted in the near future.

As Swiss scientist Claus Fröhlich explains it: "The solar constant S is the amount of total solar energy of all wavelengths received per unit time and unit area at the mean Sun-Earth distance in the absence of the Earth's atmosphere. It is customarily expressed in units of watts per square meter." The two key phrases here are "all wavelengths" and "in the absence of the Earth's atmosphere." The atmosphere must be discounted because not all wavelengths can penetrate it. Due to absorption by ozone, water vapor, and carbon dioxide, most wavelengths below 2,950 angstroms and above 2.5 microns are effectively blocked before they reach the surface of the Earth. Thus, any measurement of S made from the ground must be corrected for atmospheric effects, resulting in an inherent uncertainty for the true value of S.

A number of measurements have been made recently from far above the ground, by aircraft, balloons, and satellites. These measurements are more accurate than ground-based readings, but they also have their problems. In calibrating the measurements, one must take into account such factors as the filtering characteristics of the instrument windows and the individual quirks of the instruments themselves.

Because of all the uncertainties, there is no single, generally accepted value for S. However, the "most probable" value for the last 10 years is considered to be

$$S = 1373 \pm 20 \text{ watts /square meter}$$

Fröhlich has analyzed the data from the various types of measurements in order to bring uniformity to data taken by different groups at different times using different techniques. He finds that the mean value of S derived by high-altitude aircraft observations during the late 1960s is 1378 ± 26. For measurements made from balloons, the result is 1373 ± 14. Measurements made by three different spacecraft—Mariners 6 and 7 and the Nimbus 6 satellite —give a mean value of 1379 ± 15.

The uncertainty of these measurements is between 1 and 2 percent. Unfortunately, that is simply not good enough to tell anything very meaningful about short-term solar variations. Some observers claim to see small variations in S related to sunspot numbers, but the relationship is, in Fröhlich's words, "highly suspect." Any change in the solar constant over a time span of days to decades is likely to be extremely small. If there is any connection between such changes and terrestrial weather and climate, we cannot hope to pin it down until we have more and better data. Fröhlich writes that: "We need instrumentation accurate at the 0.1 percent level, operational for tens of years, and capable of maintaining calibration over these long periods. That challenge is surely one of the more important and most difficult problems facing experimental science."

Another way of meeting the challenge of solar variability is to look not at the overall solar constant, but rather to examine the individual components which comprise it. The vast majority of solar radiation consists of white light from the photosphere, but longer and shorter wavelengths originating in the chromosphere and the corona are also important. Radiation shorter than 3,000 angstroms is of particular interest because it has the strongest interaction with Earth's atmosphere. Solar variations in the non-visible part of the spectrum could be crucial to an understanding of weather and climate.

Spectral measurements are also difficult to make. The wide variety of experimental techniques in use and the lack of generally agreed upon absolute standards make for disagreements and controversies which are, as one scientist put it, "dismaying."

Solar radiation in wavelengths longer than 1 millimeter was discovered before World War II by amateur radio operators who noticed a persistent hiss at wavelengths of about 10 meters. The hiss appeared only during the daytime and seemed to be related to periods of intense solar activity. The wartime growth of radio and microwave technology made more detailed studies possible, and by the late 1940s scientists had discovered both long- and short-term variations in the solar radio emissions. The radio emissions consisted of three main components: quiet Sun radiation, a slowly varying signal, and brief intense bursts.

The quiet Sun radiation comes from the solar atmosphere. It is, in effect, background noise, an apparently irreducible minimum output. At meter wavelengths, the source is the corona, and at centimeter wavelengths the source is the chromosphere.

The slowly varying component, known as the S-component,

comes from the region of sunspots, plages, and intense magnetic fields. The S-component varies directly with the number of sunspots and is clearly tied to the solar cycle.

The intense radio bursts are most common during the peak of the sunspot cycle, but they can occur at any time. According to Fred I. Shimabukuro of The Aerospace Corporation, "There is a bewildering array of observed burst types. Intensities range from barely detectable levels to outputs that are orders of magnitude greater than that of the quiet Sun (at a particular wavelength) on a time scale from less than a second to days." The bursts are nonthermal in origin, and seem to result from the sudden deposition of enormous amounts of energy in the upper solar atmosphere —as in flares and transients.

In general, there seems to have been little, if any, long-term variation in the type or intensity of solar radio emissions. Since their contribution to the total solar energy flux is so small, studies of solar radio emissions are mainly important, as Shimabukuro puts it, "as diagnostic tools in identifying the solar sources of higher energy emissions."

Beyond the radio spectrum, the far infrared (FIR) emissions between 10 microns and 1 millimeter are among the most difficult to detect. At these wavelengths, the atmosphere is almost completely opaque, due to absorption by water vapor. The FIR emissions come from the coolest part of the Sun, the upper photosphere and the lower chromosphere. These emissions have almost no effect on the Earth and, like the radio emissions, their principal value to scientists is as a diagnostic tool. There seems to be some slight variation in FIR radiation with the solar cycle, but even that is minimal.

At the short end of the spectrum, solar radiation in the ultraviolet wavelengths between 1,200 and 3,000 angstroms is of considerable interest because of its impact on the Earth's atmosphere. Between 1,250 and 2,000 angstroms, the radiation is absorbed by oxygen molecules in the atmosphere, which are broken down to atomic oxygen. Radiation between 2,000 and 3,000 angstroms is absorbed in the upper stratosphere by ozone (O_3) molecules. There is now a considerable body of evidence supporting the claim that the ozone content of the atmosphere is being reduced by chemical reactions with chlorofluorocarbon molecules released by, for example, spray cans and old refrigerators. A recent study by the National Academy of Sciences warns that we may ultimately lose as much as one-sixth of our ozone in this manner. As the ozone abundance declines, the amount of UV radiation reaching the ground will increase, with presumably harmful effects for those of us who live on the ground.

Scientists have compiled tentative evidence that UV radiation varies with the solar cycle. However, most scientists are cautious about making too much of this evidence, since it is based on very few direct observations. An interesting sidelight to the question is the possibility that as the UV flux increases, emissions at longer wavelengths may decrease in some sort of compensatory mechanism which keeps the solar constant truly constant.

In the extreme ultraviolet (EUV), between 300 and 1,200 angstroms, there is solid evidence of variability, although the precise amount of variation is still open to question. The flux of EUV radiation varies in relation to the appearance of active regions, flares, and bright points on the solar disk. EUV output has also been shown to vary with the solar cycle. At the extreme short end of the spectrum, there is only a limited amount of data available on x-ray emissions. Since x-ray measurements must be taken from space, there is no long-term record to work from. Short-term variations, recorded by Skylab, are abundant, and are related to such solar phenomena as coronal holes, flares, and x-ray bright points.

Most of the Sun's energy is radiated in visible light, at wavelengths between 0.3 and 10 microns. Measurements of these wavelengths are virtually the same as measurements of the total solar constant, and are subject to the same limitations and restrictions. Not surprisingly, then, solar variation in the visible wavelengths is a highly controversial subject.

One of the leading scientists in this field is A. Keith Pierce of Kitt Peak. When I spoke with him in his Tucson office, he gave a good account of the no-nonsense approach to the subject. "You'll find people arguing blue in the face on this," he said, but Pierce clearly finds evidence more persuasive than argumentation. "The evidence," he maintains, "is that the Sun is an extremely constant source. There is no evidence of solar variation proven at the present time." In a paper written with his Kitt Peak colleague Richard G. Allen, Pierce's conclusion was only slightly less emphatic: "We thus conclude that there is as yet little or no observational evidence for any measurable change in spectral radiation with time." The difference between "no evidence" and "little or no evidence" is about as subtle as the kinds of changes the scientists are looking for. There is little to add to this, except, perhaps, the traditional scientific caveat that the absence of evidence is not necessarily evidence of absence. The question remains open.

There is probably no neat and simple way to sum up the evidence—or the lack of it—for solar variability. The field is rife with controversy, and the position one takes may ultimately depend not on the evidence, but rather on one's psychological disposition toward stability versus change. Perhaps the only safe thing to say is that the most changeable thing about the Sun is our attitude toward it. The comfortable verities of only a few years ago have been shaken and shattered, and no one really knows what the Sun will do tomorrow—except that we can predict with some confidence that it *will* rise in the east.

Postscript. Early in 1980, NASA launched the Solar Maximum Mission —known as Solar Max—in hopes of learning more about the Sun's behavior during the peak of the sunspot cycle. The Earth-orbiting satellite has already returned a bounty of intriguing new data. Experimenter Richard Willson, of the Jet Propulsion Laboratory, reports evidence of significant short-term solar variability. Twice during the spring of 1980, the solar output dipped by about 0.2 percent for periods as long as a week. This decline in the solar constant coincided with the appearance of unusually large sunspot groups. Shorter-term variations, ranging from 0.01 to 0.1 percent, seem to be taking place virtually all the time.

The Solar Max results are important, both for what they tell us about the Sun, and because of their implications for the Earth. Willson suggests that a drop in solar output of 0.2 percent could be enough to have a direct effect on terrestrial weather. For a discussion of other possible solar-weather links, see Chapter 8.

Seven: FIRE AND ICE: THE SUN AND CLIMATE

The city fathers of Miami have decided to put in a bid for the next Winter Olympics. Their chances are slim because of the lack of nearby mountains, but something must be done to revive the dying economy. Honolulu and Singapore have also put in bids, but the odds are that the Olympics won't be held, anyway. Life is getting too grim for games. The Mexican wheat crop was disappointing, and imports from Brazil are dropping; the shipping lanes are clogged with icebergs. The latest report from United Nations headquarters in Rio forecasts a continued shortening of the growing season. Heating fuel will be in short supply because the glaciers have finally covered the Pennsylvania coal fields. The laser experiments in Denver have failed completely, and the city will soon be crushed by the ice surging down from the Rockies. Meanwhile, the Saharan monsoons continue, and the reclaimed desert farmland is turning to a quagmire. No one knows what the situation is in Calcutta, where it hasn't rained in three years. The huge grain harvests from Chad and Nigeria might have relieved the spreading famine, but the sporadic nuclear war with South Africa has contaminated much of the crop. It is a bitter time for everyone, except, perhaps, for the Stone Age tribesmen on a remote Philippine island, who are excited by the strange, fluffy white stuff floating down from the sky. They have never seen it before; perhaps they should worship it.

The noonday Sun shimmers in the sky like a slab of melting butter. The buckling, bubbling streets are deserted. The few Christmas shoppers have made their rounds early and retired to whatever shade they can find. Even the scavengers and street mobs are in hiding, waiting for the clammy comfort of dusk. Manhattan is a frying pan. Smoke from the fires in Jersey drifts across the rising river on a hot wind. In the dense shadows of apartment houses and tenements the stink of rot and decay is overwhelming. There is no air conditioning, no refrigeration, no electricity; the ban on burning of fossil fuels has left a gap that cannot be filled by the scattered solar energy systems. Less than a million people remain in New York City. The rest have died or taken to the roads leading north, only to be turned back at the Canadian border by armed patrols. For those who remain, the waiting will soon be at an end. Already, the waters are lapping at the sandbags on Wall Street. New Orleans has long since been evacuated, as have Houston, Rotterdam, and Venice. It has been months since the last report from the tropics. Malaria is running rampant in Montreal, and bubonic plague has claimed thousands in Los Angeles. World population is now less than a billion, and millions more are dying each day from starvation. Food supplies will not last until spring, but by then it won't matter. By next summer, the oceans will boil. After four billion years, life on the planet Earth is about to become extinct.

Scenarios for the next Irwin Allen disaster flick? Possibly. It is also possible that these are scenarios for life (or its demise) in

the twenty-first century. Of all the morbid predictions of the apoc-
alypse, climatic change is the only one which is certain to come
to pass. We just don't know how or when.

We know that the Sun plays a crucial role in determining the
Earth's climate, but we are only beginning to understand the
mechanisms involved in climatic change. There is reason to believe that we are on the verge of another ice age; but there is also
evidence that our own mistakes and carelessness will soon turn
the world into an uninhabitable hothouse. Between these two ex-
tremes there is a broad spectrum of opinion about what the future may hold.

The key to the future may be found in the past. From the fossil
record, we know that 100 million years ago there were steaming
tropical swamps in the Antarctic. Just 20,000 years ago the fertile
plains of the Midwest were buried under mile-high glaciers. A
mere 2,000 years ago, the Carthaginians flourished in what is now
a dry and sterile desert. And today, we watch helplessly as the
Sahara marches south, displacing entire populations.

Climatic change may be caused by dozens of different mechanisms, working alone or in combination. But whatever factors may
be involved, climate begins and ends with the Sun. The amount
of solar energy reaching the surface of the Earth—known as insolation—is the fundamental control on terrestrial temperatures.
One reason the search for solar variations is so important is that
man cannot hope to understand Earth's climate until the basic
parameters of insolation are established.

Unfortunately, at the present time the keystones of solar theory
and climate theory seem to be incompatible. If the Sun works the
way we think it ought to, the Earth should be a frozen wasteland.

THE ANCIENT SUN AND THE ANCIENT EARTH

The major stumbling block is the apparently well-established
theory of the faint early Sun. From solar models and observations
of other Main Sequence stars, astronomers are convinced that during its early history, the Sun must have been some 30 percent less
luminous than it is today. Over the past 4.5 billion years, the Sun
has gradually become more luminous. According to NCAR scientist Gordon Newkirk, "The mechanism that produces that
[change] is one which seems pretty inescapable in terms of what
we think we know about how the Sun gets its energy."

The standard theory holds that the Sun has increased its luminosity at a fairly uniform rate over the last 4.5 billion years. As hydrogen is consumed, the helium "ashes" collect in a shell around the central core. Because helium is slightly heavier than hydrogen, the core becomes compressed, resulting in a steady increase in pressure, temperature, and energy production.

Periodically, this steady increase may be broken by abrupt changes in luminosity, amounting to as much as 30 percent. The helium shell around the hydrogen core is unstable. When the shell reaches a critical size, the helium begins to mix downward into the core. This touches off a sudden increase in energy production in the core, coupled with a complementary drop in surface luminosity. Eventually, the effects of the core events are felt on the surface, and luminosity rises again. When enough of the helium shell has been consumed, balance is restored and the Sun returns to its slow and steady climb in luminosity.

Intriguingly, the calculated interval between these episodes of helium burning is about 300 million years. This corresponds to the typical time scale of terrestrial epochs of glaciation. It also is about the same length as the galactic year. Thus, there are at least two suspected mechanisms to account for periodic ice ages on Earth. What is lacking is a solid explanation for why the Earth didn't become an ice world during the early history of the Sun.

As mentioned earlier, planetary scientists believe that even a 10-percent drop in solar luminosity would result in runaway, irreversible glaciation on Earth. One can tinker with the models and get different figures, ranging from 5 percent to 30 percent, but the outcome is always the same. A cooler Sun means bigger polar caps, more ice and snow, and more energy reflected back into space, resulting in still lower temperatures. Eventually, the Earth falls into what Newkirk refers to as "a permanent thermodynamic hole."

Yet the geological record indicates that this glacial catastrophe never happened. Whatever the climate may have been doing billions of years ago, there is no evidence that the mean global temperature ever fell below the freezing point of water. Somehow, the Earth avoided becoming another Mars.

There is a way out of this dilemma. The Sun has changed over the course of 4.5 billion years, but so has the Earth. There is abundant evidence that the young Earth's atmosphere was considerably different from the present day oxygen-rich atmosphere.

Before about 1 billion years ago, there was very little oxygen. The presence of unoxidized iron and lead deposits and the existence of anaerobic microbes indicate that Earth's early atmosphere was oxygen-poor and hydrogen-rich. The primitive atmosphere may have been similar to the atmospheres of Jupiter and Saturn, where compounds such as methane and ammonia are abundant.

An early atmosphere rich in methane, ammonia, and carbon dioxide would not have been so sensitive to the faint early Sun. These gases are excellent absorbers of infrared radiation. Instead of bouncing back into space, much of the Sun's heat would have been trapped by the carbon compounds in the atmosphere. This is known as the greenhouse effect, and it almost certainly accounts for Earth's escaping runaway glaciation.

But there is also a runaway greenhouse effect, which is probably responsible for the sultry conditions on Venus. If the greenhouse kept Earth from becoming another Mars, what kept it from becoming another Venus?

Caught between fire and ice, atmospheric scientists have been forced to tread a very thin and slippery line. A variety of proposals have been put forward to explain why Earth is such a nice place to live, but no single theory seems adequate to explain everything. Newkirk is one of the leading scientists in this field, but he is quick to point out that "all the ideas about the evolution of the atmosphere of the Earth are pretty primitive compared to our ideas about the evolution of, say, a star. This is largely because the whole question of the evolution of an atmosphere is very, very, very complex."

Basically, the standard theories hold that the early hydrogen-rich atmosphere gradually dissolved out in the oceans, or escaped to space, and was replaced by biologically produced oxygen. Meanwhile, the Sun was getting warmer, and, by the time the greenhouse disappeared, it was no longer needed. "But," says Newkirk, "there's an aspect of it that worries me . . . and quite a few other people. Unless [you] start out with just *exactly* the right conditions, the atmosphere goes unstable. It either experiences the runaway glaciation or the runaway greenhouse effect. . . . So you come to the conclusion that the real atmosphere was more robust than these models are." Newkirk, like any good scientist, is suspicious of answers which seem to rely on a stacked deck. "It sounds a little bit strange," he says, "that something is so, so sensitive that it basically is unstable. If you pick just a little bit different abundance for, say, carbon dioxide, or a different reaction rate for carbon dioxide being dissolved in water, then the thing falls off into one of these cataclysms."

It would seem that we lucked out. But such a narrow escape is a little too close for scientific comfort. All of this, says Jack Eddy, "drives you to an interesting argument that says, 'Gee, it looks as though the Earth's atmosphere kind of adjusts to keep things nice for us.' If that's the case, then you're into a nice philosophical situation. Is there suddenly proof that there's a God?"

Theological arguments aside, there may be a sound scientific reason for our good luck. At this point in our conversations, Eddy, Newkirk, and several other scientists referred me to what is known as the Gaia Hypothesis. This is a new and somewhat exotic theory which suggests that the Earth's atmosphere may behave like a huge living organism.

The main proponents of the theory are James Lovelock, a British atmospheric chemist, and Lynn Margulis, a microbiologist at Boston University. The hypothesis takes its name from the ancient Greek earth goddess in whose body all living things were organs. It has semimystical overtones which are troublesome, but an increasing number of scientists seem to be taking it seriously.

Lovelock and Margulis look at the Earth's atmosphere and see a number of phenomena which ought not to be there. Aside from the problem of glaciers versus greenhouse, they point to such things as the abundance of atmospheric oxygen, which ought to result in highly acidic soil but doesn't, and the fact that the continents (via erosion) should be depleted in essential trace elements but aren't. They suggest that: "There are, in the atmosphere, 'too many' of exactly those elements that are required by living things." The key factor in maintaining a biologically favorable climate, they say, is the presence of living organisms.

The Gaia Hypothesis maintains that terrestrial ecology is, in effect, self-correcting. Whenever something unbalances the climate, a host of biological mechanisms go to work to set things right again. One need not invoke a "guiding hand" to keep the system running, although some people believe that it would help. In any case, biological feedback serves to modulate the unbalancing influence of such external factors as solar variation and volcanic eruptions. Lifeless Venus gets no such help. The verdict on Martian life is still not in, although Viking results seem to indicate the absence of living organisms—an outcome which was predicted by Lovelock and Margulis.

We will return to the Gaia Hypothesis in a different context in Chapter 9. For the moment, it does seem to provide a possible solution to the problem of the faint early Sun and terrestrial climate. But, for all its many subtleties, the Gaia argument is, in some ways, circular. In the end, it seems to say that life exists on Earth because life exists on Earth. We're here because we're here.

CLIMATE CYCLES

Long before we were here, primitive organisms were leaving a fossil record in the geological strata of the young Earth. Scientists have now traced life back for 3.8 billion years, although the fossil history is very incomplete before about 1 billion years ago. By analyzing the types, locations, abundances, and isotopic composition of the ancient fossils, we can draw some very broad generalizations about the early climate. Since life seems to have been here continuously for nearly 4 billion years, we can conclude that during that period the mean global temperature must have been between the freezing and boiling points of water. There seems to be no evidence for solar variations during this period exceeding 10 to 20 percent—although the problems with atmosphere and the faint early Sun cast some large doubts on the significance of such a conclusion.

The record is much clearer for the last billion years. By this time, life had evolved sufficiently to leave widespread macroscopic fossils which give an indication of prevailing climatic conditions. The evidence points toward three major episodes of extensive glaciation, occurring 700 million, 300 million, and 10 million years before the present time. Over the last 100 million years, there seems to have been an episode of rising temperatures between 90 million and 75 million years ago, followed by gradual cooling which culminated in the most recent ice age.

For the past million years, the global climate seems to have been dominated by successive episodes of advancing and retreating glaciation, with periods of about 100,000 years, 41,000 years, and 23,000 years. The last great wave of ice retreated between 14,000 and 10,000 years ago. Since then, there seems to have been a 2,500-year cycle of cooling, as evidenced by the advance of mountain glaciers. The most recent glacial episodes occurred between 3800 and 2900 B.C., 1300 and 400 B.C., and 1300 and 1900 A.D. From the late nineteenth century through about 1940, global temperatures rose slightly; in recent decades, the trend has apparently been downward.

It would be convenient if we could simply eliminate all other factors and concentrate on the Sun's role in these climatic excursions. Unfortunately, a bewildering array of factors, of which the Sun is only one, contribute to climatic change. "I get the impression," says Newkirk, "there are more theories of ice ages than the world needs right now."

Probably the major factor in the 300-million-year cycle of ice ages is continental drift. The continents have been skidding around on huge crustal plates, colliding with one another, forming immense supercontinents, then breaking up again. Obviously, the placement of continents has a large influence on ocean currents, which are one of the important elements in determining climate. A critical factor seems to be the location of the North and South poles. If the poles are centered on an ocean, warm tropical water can circulate to high latitudes and keep temperatures moderate. If the poles are over continents, this circulation is prevented and permanent ice caps form. Three hundred million years ago, for example, the South Pole was located in what is now the continent of Africa. Today, of course, the South Pole is located in the antarctic continent. The North Pole is over an ocean, but an ocean so completely surrounded by landmasses that the effect is virtually the same as if the pole were located on a continent.

The 100,000-, 41,000-, and 23,000-year glacial cycles are almost certainly related to defects in the Earth's orbit around the Sun. The Earth revolves, not around the center of the Sun, but around the center of all the mass in the solar system. The Sun contains 98 percent of that mass, and most of the remainder is Jupiter. The influence of the mass not contained in the Sun moves the focus of the Earth's orbit slightly offcenter with respect to the Sun. This results in periodic changes in the shape of Earth's orbit.

On a cycle of about 100,000 years, the Earth's orbit varies from nearly circular to somewhat elliptical. The 41,000-year cycle involves an up-and-down "nodding" of the Earth's rotational axis with respect to the Sun. The 23,000-year cycle is caused by the effect of lunar and solar gravity on Earth's slight equatorial bulge; the result is a "wobble" in the axis of rotation which causes the poles to wander in a broad circle.

Although none of these cycles changes the average Earth-Sun distance, their existence does change the angle and the amount of sunlight reaching the Northern and Southern hemispheres during a given season. When the cumulative effect of these orbital and rotational variations is cool northern summers and very cold southern winters, the result is an ice age. In the Northern Hemisphere, less polar ice melts during the cool summer; in the Southern Hemisphere, more sea water freezes during the harsh winter. As the total amount of ice increases, more solar radiation is reflected back into space, and the global temperature drops.

These cycles were first identified by the Yugoslavian scientist Milutin Milankovich in the 1930s. The Milankovich model was hotly disputed for years because glacial episodes that matched all of the cycles could not be found. But in recent years the accumulation of better historical data seems to have confirmed the existence of Milankovich's cycles.

The Sun thus seems to be off the hook as far as major ice ages are concerned. We don't need solar variations to account for the great epochs of glaciation. But, again, a caution is necessary. Before the neutrino controversy arose, it was an article of faith that the Sun never changes very much. Given that limitation, geologists studying ice ages naturally preferred to concentrate on nonsolar explanations. Returning to our Murder at the Superbowl mini-drama, if a prime suspect seemed to have an airtight alibi placing him in Istanbul at the time of the murder, the detectives would turn their attention to other suspects. But if the alibi begins to come unraveled, then one has to consider the possibility that this suspect was at the scene of the crime, after all.

If the Sun has beaten the rap for long-term climate variations, it remains a strong candidate for the source of short-term changes. The best evidence for solar-induced climate swings comes from tree rings—although not in the way one might expect. Although A. E. Douglass found a correspondence between tree ring growth and the Maunder Minimum, this was really only a coincidence. If you go outside and chop down the nearest tree, you will be unlikely to find any hint of an 11-year cycle. Tree ring growth depends on the amount of local rainfall, which has no demonstrated connection with the Sun. What you will find, however, is varying amounts of carbon-14 in the cells of the tree.

Carbon-14 is produced in the upper atmosphere by the collision of cosmic-ray particles with nitrogen-14 atoms. The more cosmic rays, the more carbon-14. As mentioned earlier, the cosmic-ray flux depends in large measure on the level of solar activity. At times of high solar activity (sunspots, flares, and so forth), the Sun's influence shields the Earth from the incoming galactic particles; when solar activity is low, more cosmic rays reach Earth's atmosphere. Thus, there is an inverse relationship between solar activity and carbon-14 abundance.

Plants assimilate atmospheric CO_2 into their cells. Most of the carbon in these CO_2 molecules is stable carbon-12. By measuring the ratio between the slightly radioactive carbon-14 and the stable carbon-12 content of tree rings, we can calculate the amount of carbon-14 in the atmosphere at the time the tree was growing. After adding on the fifty years it takes for the carbon-14 to get from the upper atmosphere to the plants, we should have a fairly precise indicator of the level of solar activity for a particular era.

Eddy has found a strong match-up between carbon-14 abundance and the Maunder Minimum, which corresponds with the so-called Little Ice Age of the seventeenth century. During that period, carbon-14 levels were enhanced by about ten parts per million. Although other factors (chiefly the strength of Earth's own magnetic field) can influence carbon-14 production rates, the link between solar activity and carbon-14 seems well established.

What is still missing, however, is an ironclad link between changes in solar activity and changes in climate. Statistically, the relationship between carbon-14 excursions and episodes of glaciation over the last 7,000 years is highly suggestive. But it is not enough to say that if A and B occurred at the same time, A therefore must have caused B. Some physical mechanism must be identified which establishes a causal relationship between A and B. That missing link is yet to be demonstrated.

The most obvious mechanism would be a slight change in solar luminosity related to the sunspot cycle. There are data which suggest that such a change does occur, but as mentioned in Chapter 6, the difficulties involved in determining the solar constant make suspected variations of less than 2 percent highly uncertain. There is some evidence that changes at the level of 0.4 percent may occur over the course of several years, but there are no firm data pointing to a cyclical change in the value of S related to the sunspot cycle.

Even if the solar constant does change, climate models must account for a terrestrial response to that change. That is not easily done. "The Earth's atmosphere has a lot of inertia," Eddy points out. "You don't just kick it and make it rain someplace. It's amazingly resilient. You turn off the Sun at night and things don't freeze up. You go from summer to winter and things sort of even out. . . . If the Sun were to turn off momentarily, or even for a day or two, the Earth, I think, soon would get over it." To affect climate "a persistent, nagging" change in the Sun would be needed.

Leaving aside for the moment the details of atmospheric response to the Sun (which are not well understood), one can make a moderately strong statistical argument that some sort of terrestrial response does occur, roughly in phase with the 22-year solar cycle. Of course, if one plays with numbers long enough, correlations are inevitable (for example, my sunspot—stolen-base connection). The value of such statistical arguments is that if a real correlation can be demonstrated, then the search for connecting mechanisms acquires better direction and focus.

The best evidence for a solar-related climatic cycle seems to be the 22-year recurrence of droughts in the western United States. The Dust Bowl, *Grapes of Wrath* episode of the 1930s is well known. In the early 1950s, another drought hit the region. And right on schedule, a third episode occurred in the mid-1970s. There is good evidence that this cycle extends at least 300 years into the past. One might reasonably expect to see another drought in the late 1990s.

The problem here is that the solar cycle does not selectively attack the western United States. If the drought cycle is truly related to the Sun, then global effects should be evident. It would be a monumental task to comb the meteorological archives of the entire world (such as they are) and produce a meaningful correlation between widely separated events. Without a systematic analysis of global data, the North American drought cycle can only be viewed as a strong suggestion of a solar link, and not as conclusive evidence.

In terms of climate prediction, the 22-year drought cycle isn't even very helpful to those who live in the affected region. During the Dust Bowl era, rainfall was plentiful in Ontario and Saskatchewan. The 1977 drought which had Californians placing bricks in their toilet tanks to conserve water wasn't felt in Iowa, where there were floods. In a very general way, the data is meaningful to those searching for solar-climate links, but it is not very impressive to those who doubt the existence of such a connection. "This sort of thing," one scientist told me, "makes classical climatologists yawn."

Given our uncertainty about what has happened in the recent past, it is almost impossible to say anything definite about what the future will bring. That, of course, does not prevent people from making predictions anyway. After the California drought and the extremely harsh winters in eastern North America in 1977 and 1978, a lot of people were heard to say, "The weather is getting weird!" But the weather is *always* weird—somewhere. If the unusual conditions occur in Tasmania or Afghanistan, nobody (except for Tasmanians and Afghans) really notices. If, on the other hand, the weather is extreme in Chicago, Boston, or London, *National Enquirer*-style headlines begin to appear, announcing the onset of another ice age, or the beginning of a runaway greenhouse effect. Sensational articles and paperback books presenting doomsday scenarios—such as those at the beginning of this chapter—have been increasingly common in recent years. The proliferation of such tracts, including many which are merely astrology and numerology masquerading as "science," may have obscured the fact that real and potentially devastating climatic changes may actually be taking place.

THE CHANGING CLIMATE

One must approach the subject with extreme caution. Hedging is permitted, and even encouraged. If the climate is changing, crying "Wolf!" will not help the situation. An ice age is a real possibility, as is a greenhouse situation. Since the two are mutually exclusive, the evidence for each must be looked at carefully.

The reality of the Milankovich cycles is the most telling evidence for another ice age. Another episode of glaciation seems inescapable; indeed, by some definitions, we are already in an ice age, and have been throughout the entire history of human civilization. If we look at the history of the last 10 million years, interglacial episodes seem to be the exception, rather than the rule. The return of the glaciers is apparently inevitable, and the only question is when it will happen.

There are two general theories of how ice ages begin. Both, or neither, may be correct. The slow-freeze theory holds that the gradual build-up of snow and ice due to cold winters and cool summers takes hundreds to thousands of years, eventually causing an albedo increase and a decline in temperatures. Year by year, the glaciers creep outward from the poles and down from the mountains.

A competing theory suggests that at least some glacial episodes may happen with frightening speed. In what is known as a "snowblitz," just a few cold years in succession would be enough to unleash the glaciers. There is some geological and paleontological evidence that snowblitzes have happened in the past, preserving for us the frozen carcasses of some presumably slow-footed or unlucky woolly mammoths. A snowblitz ice age could conceivably happen in less than a hundred years; our grandchildren could see a world in which Miami has a climate comparable to that of present-day New York.

Since 1950, the mean temperature in the Northern Hemisphere has apparently declined slightly. Growing seasons in some areas of Canada and Europe have shortened. It is too soon to tell whether this is evidence of a long-term trend or merely a temporary fluctuation. Glacier-watchers report advances in some regions, retreats in others. The ambiguities are many. It would seem that at this point, forecasts of an impending ice age are probably no more reliable than the annual predictions of the *Old Farmer's Almanac*.

If an ice age doesn't happen, mankind may be responsible. That would not, however, necessarily be a cause for self-congratulation. The activities of modern man may be leading straight into a global greenhouse.

As with ice ages, the evidence for global heating is confusing, contradictory, controversial, and not at all conclusive. The leading spokesman for the greenhouse view is probably Stephen H. Schneider of NCAR. Schneider is often described as "the Carl Sagan of atmospheric science," a comparison neither man would be likely to resent. Schneider is young, articulate, socially concerned, and not afraid to go out on a limb. Like Sagan, Schneider is not reluctant to come down from the scientific mountain (in his case, literally; NCAR is located on a mountain overlooking Boulder, Colo.) and explain his work to the general public. He has co-authored a popular book, *The Genesis Strategy*, expressing his concerns about what we may be doing to the global climate.

According to Schneider, mankind has been pumping carbon dioxide into the atmosphere at a potentially dangerous rate. As the CO_2 level rises, less solar heat escapes and the mean temperature goes up. If the present trend continues, the average global temperature could increase by 1°C by the end of the twentieth century and 2 to 3°C by the middle of the twenty-first. "These seemingly insignificant changes," Schneider argues, "are sufficient to disrupt the earthly heat and water balance."

Carbon dioxide (CO_2) gets into the atmosphere naturally through a variety of chemical, geological, and biological processes. The ocean, in turn, removes CO_2 by dissolving the gas and forming carbonate (CO_3) deposits. As long as the rates involved remain relatively constant, the system is stable. But ever since the Industrial Revolution, man has been burning fossil fuels and depositing ever-increasing amounts of CO_2 in the atmosphere. The level of atmospheric CO_2 before the Industrial Revolution is thought to have been about 265 to 290 parts per million; the current level is about 330 ppm. By the turn of the century, the CO_2 level will be around 400 ppm. Before the middle of the twenty-first century, the figure could be above 500 ppm. The ocean simply cannot deal with the increase.

The rising temperatures associated with CO_2 pollution may moderate the natural trend toward an ice age. But the direct climatic response to the heating would be subtle and unpredictable. Ocean currents and prevailing winds could alter, bringing droughts and monsoons to regions where these don't occur today. A more dramatic effect could be the melting of the polar ice caps. Some scientists are more than a little concerned about the fate of the West Antarctic ice pack, which seems to be resting on a layer of slush. Even a slight increase in temperature might be enough to cause the glaciers to slide into the ocean.

If the ice caps should melt significantly, the sea level would rise by 15 to 25 feet. Even if we had time to get out of the way of the rising waters, the effect of such an event would be catastrophic. In analyzing the consequences of a 25-foot rise in sea level, Schneider of NCAR and Robert Chen of the Massachusetts Institute of Technology, together found that in the United States alone, some 15 million people would be rendered homeless, and 150 billion dollars (1971 dollars) in property would be lost. Some areas would be utterly devastated; Florida, for example, would lose about 50 percent of its population and 60 percent of its wealth.

We don't know that this will happen. By the time we can be sure, it may be too late. According to Schneider, "It is unlikely that scientists will provide reasonable certainty of the magnitude and timing of potential climate modifications before the atmosphere 'performs the experiment' itself."

If there is any cheerful news in all of this, it is that Schneider doubts that the greenhouse would reach the irreversible, runaway Venus case within about 100,000 years. Still, the effects of climatic change cannot be ignored. The prudent man, Schneider argues in *The Genesis Strategy*, should hedge his bets and make plans for the "lean years" which may await us.

With or without the greenhouse or the snowblitz, it is clear that the climate is going to change. Our idea of a "normal" climate is badly distorted. According to Reid Bryson, of the University of Wisconsin, the international climate "norms" established during the period from 1930 to 1960 are actually based on what was the most abnormally warm period in the last 1,000 years.

Our international economy is based on those abnormal norms. The interdependence of the modern world makes it impossible to isolate the effects of even minor climate alterations. If the monsoon rains in India are late, farmers in Kansas and Alberta will have to feed millions of additional people. But if climate change hits the traditional breadbasket regions, such as the American Midwest and the Ukraine, the global effects will be catastrophic. Even with the optimal conditions and bumper crops of recent decades, a billion people—a quarter of the world's population—are suffering from malnutrition. We have seen the horrible results of politically induced starvation in Biafra and Cambodia; if the climate changes, those tragedies will be multiplied many times over.

It is difficult to plan for such a future. To politicians worried about next year's election, a famine in the next century is not a high-priority concern. A farmer who wants a good price for his wheat crop this year is not likely to grow a surplus for the sake of those who will be hungry in some hypothetical future. If history is any guide, we will probably not even acknowledge the problem until it is already upon us—as the oil shortage, pollution, nuclear hazards, and overpopulation have already demonstrated.

The Sun's role in all of this is problematical. We are only beginning to understand the links between the Sun and climate. In the short term, it may not even matter if the Sun's output is variable; other factors are probably more important. Yet ignoring the Sun's contribution to climate would be foolish. To appreciate our absolute dependence on the Sun, we have only to look at our nearest neighbors in space, Mars and Venus, and consider the delicate balance which separates those worlds from ours. In the end, it's not just a small world, it's a small solar system.

Eight: SUNNY WEATHER

One of the great ironies of modern science is that we can predict with some confidence the arrival of another ice age a thousand or more years from now, but not whether it will rain tomorrow. As I write this, the city of Buffalo is digging out of an unpredicted 27-inch snowfall, while the streets of Boston remain dry and clear since an expected 4 inches of snow never fell, to the dismay of skiers and the delight of the rest of us. Despite the advent of radar, computers, and satellites, meteorology is still a Delphic art, part science and part luck. The accuracy of weather forecasting has undeniably improved in recent years, but our confidence is based more on general trends than specific events.

The Sun is the engine which drives what British science writer Nigel Calder refers to as "the weather machine." Solar radiation provides the energy for breezes and tornadoes, showers and monsoons, Canadian highs and tropical depressions. But despite the Sun's central role, very few meteorologists have been inclined to include solar influence in their calculations of day-to-day weather patterns. Although it may seem logical that there should be direct connections between the Sun and terrestrial weather, it has been extremely difficult to prove the existence of such links.

THE EARTH'S ATMOSPHERE

Lately, however, scientists are beginning to understand the workings of some of the many gears and levers in the weather machine. Our knowledge of the atmosphere is far from complete, and the links between it and the Sun remain mysterious, but we are at last approaching the point where, if we can't *do* anything about the weather, we can at least talk about it more meaningfully.

The first thing to understand about the Earth's atmosphere is how little of it there is. The total mass of the Earth's atmosphere is about 5.8 billion tons, which is roughly one-billionth of the mass of the planet itself. The equatorial diameter of the Earth is 12,756 kilometers; the breathable portion of the atmosphere extends that figure by only 10 kilometers. If you scale that down to the size of a standard 12-inch desk globe, the top of the troposphere would be less than 5/1,000 inch above the surface of the globe. The upper atmosphere, including the thinnest reaches of the ionosphere, goes out to about 500 kilometers above the surface; on our desk globe, that would be an altitude of slightly less than ½ inch.

Such statistics are meaningful. They show that Earth's atmosphere is thin and diffuse and that what we think of as weather occurs in only the very bottom layer of it. Such a thin envelope should be extremely sensitive to minor changes in the Earth or the Sun. On the other hand, the Sun is 93 million miles away (or about 3.3 miles from our desk globe). One would expect, then, that terrestrial influences would have a far greater effect on the atmosphere than would anything associated with the distant Sun. Although heat from the Sun drives the entire weather system, such earthly factors as mountain ranges and ocean currents would

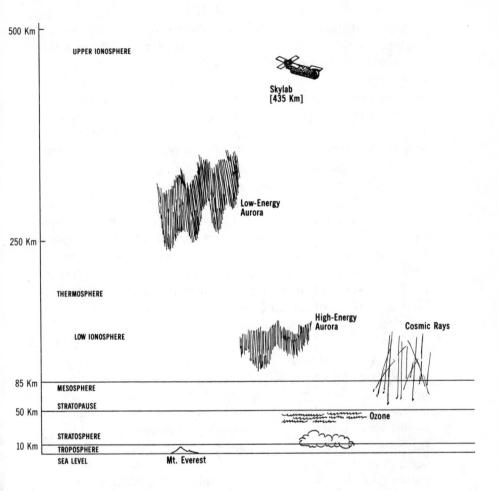

500 Km

UPPER IONOSPHERE

Skylab
[435 Km]

Low-Energy
Aurora

250 Km

THERMOSPHERE

LOW IONOSPHERE

High-Energy
Aurora   Cosmic Rays

85 Km   MESOSPHERE

STRATOPAUSE

50 Km

Ozone

STRATOSPHERE

10 Km   TROPOSPHERE

SEA LEVEL   Mt. Everest

The Earth's atmosphere acts as a series of filters, preventing most wavelengths of solar radiation from reaching the ground. Subatomic particles from flares and the solar wind interact with the thin atmosphere at high altitudes, creating auroral displays. Lower, the ozone layer absorbs most of the harmful ultraviolet radiation. What we think of as weather occurs only in the very lowest layer of the atmosphere and may or may not be directly affected by variations in the solar output.

seem to be far more important in controlling day-to-day weather patterns. That is essentially the position taken by classical meteorologists, who see no reason to invoke solar influences when trying to predict tomorrow's weather.

There is, however, a small but growing body of evidence which suggests that minor, short-time-scale solar variations may have at least an indirect effect on global weather patterns. This is a subject which is hotly disputed, and a final verdict seems remote. To date, most of the evidence for a Sun-weather link is statistical and not supported by proven physical mechanisms which could account for the suspected correlations. Atmospheric scientists are now concentrating on a search for such mechanisms, because, as Stephen Schneider puts it, "statistical correlations in the absence of causal mechanisms give little physical insight or understanding of the solar-terrestrial system."

The Earth's atmosphere acts as a filter, or a series of filters, shielding the surface from most wavelengths of solar radiation. Because different wavelengths are absorbed at different altitudes, a wide and confusing array of physical processes is at work in the various atmospheric layers. Complex and subtle details of chemistry and physics are involved, but it may be useful to attempt a brief description of some of the interactions which seem to have an effect on weather conditions.

The Earth first feels the Sun's influence at the magnetopause: the boundary between Earth's magnetic field (the magnetosphere) and the interplanetary medium. The magnetosphere is not really a sphere, but has a shape more like that of a teardrop, and is constantly changing in both shape and size. An indication of just how variable the magnetosphere can be was provided by the Pioneer and Voyager missions to Jupiter. Befitting its grandiosity, Jupiter has an immense magnetosphere, which some scientists describe as the largest structure in the solar system. Its influence can even be felt by the Earth. When Pioneer 10 first passed the Jovian magnetopause, scientists assumed that the spacecraft had seen the last of the solar wind, which is deflected by planetary magnetic fields. They were surprised, then, when they observed Pioneer 10 crossing the magnetopause six more times. The spacecraft was not making U-turns. Rather, the strength of the solar wind was varying. As its intensity increased, it compressed the Jovian magnetic field, pushing it inward toward the planet and past the incoming probe. As the magnetosphere pulsated in response to the varying solar pressure, Pioneer 10 crossed and recrossed the boundary. The Earth's much smaller magnetosphere is similarly manhandled by the solar wind.

The magnetosphere extends outward from the Earth for about

ten Earth radii (around 60,000 kilometers) in the direction of the Sun; away from the Sun, it trails off in an extended cometlike tail. The geomagnetic field lines which form the magnetosphere emerge from the Earth at high latitudes near the magnetic poles —again, a simple bar magnet provides a useful analogy. High-energy solar particles spiral inward along these field lines to reach the upper atmosphere, where they collide with the molecules of atmospheric gases.

THE AURORA

These high-energy collisions produce the auroral displays known as the Northern and Southern lights. About half of the energy of the incoming particles is expended in exciting oxygen molecules at an altitude of around 90 kilometers. The result is an eerie, green, translucent glow. Lower-energy solar particles cannot penetrate that far, but they produce a similar effect on free oxygen atoms at an altitude of about 250 kilometers. There, at a lower energy level, the glow is red, rather than green. A mixture of high- and low-energy particles can result in an awesome multicolored light show.

Most of us seldom get to see such displays because they are generally confined to high latitudes. When the solar wind is augmented by intense flare activity, however, the magnetosphere can become so badly distorted that the field lines are overwhelmed and the auroras spread to the middle latitudes. There have even been reports of auroral displays visible in Singapore, on the equator.

A tremendous amount of energy is involved in the auroras. Some of the energy transferred from the solar wind is temporarily stored in the magnetosphere. When enough energy has built up, instabilities arise, and the excess energy is suddenly discharged into the atmosphere. This event is known as a geomagnetic substorm. The substorms occur about five or six times every day, and endure for a few hours. Typically, the daily energy input from substorms is on the order of 10^{12} watts.

During periods of high solar activity, large-scale geomagnetic storms can occur, in addition to the more predictable substorms. The big storms are ten times as energetic as the substorms and may last for several days. The infusion of high-energy particles increases the ionization of the upper atmosphere and may have a disruptive effect on radio communications.

Another effect of increased solar activity is the heating of the upper atmosphere at an altitude of about 250 kilometers. This inflates the atmosphere, pushing it into regions where it is normally negligible. Skylab was a victim of this inflation—as were the NASA engineers who didn't expect it to fall so soon. Higher than normal solar activity increased the air drag on the orbiting laboratory, causing its orbit to decay ahead of schedule. NASA's embarrassment over the incident was ironic, in light of the fact that Skylab's primary object of study was the Sun.

What happens to the energy injected by the auroras is a subject of extreme interest. One of the leading scientists in the field is Raymond G. Roble of NCAR. Roble suspects that the energy becomes involved in a variety of upper atmospheric reactions. "We would be extremely surprised," he says, "if it came all the way to the ground." Although there is good statistical evidence that auroral displays vary in intensity with the solar cycle, tying the auroral statistics to the weather statistics will be a problem. Roble writes that "it is going to be very difficult to find and verify auroral effects on the lower atmosphere, particularly those associated with weather events."

THE THERMOSPHERE AND THE STRATOSPHERE

The dynamics of the upper atmosphere between about 80 and 500 kilometers are tremendously complex. The most important energy source in this region, known as the thermosphere, is solar radiation in the extreme ultraviolet at wavelengths between 300 and 1,250 angstroms. About half the energy goes into the auroras, and the rest cascades downward through a series of chemical and kinetic reactions. The thermosphere is difficult to understand because it is strongly affected by both the solar wind and the absorption of electromagnetic solar radiation. In effect, in the push-pull war between the Sun and the Earth, the thermosphere is the battleground.

Between 10 and 85 kilometers are the stratosphere and the mesosphere, separated by the ozone layer at an altitude of about 50 kilometers. Here, solar radiation is responsible for some complicated and important chemistry. Ultraviolet radiation between

1,800 and 3,100 angstroms drives a cycle of reactions involving molecular oxygen (O_2), free oxygen (O), ozone (O_3), nitrous oxide (N_2O), nitric oxide (NO), methane (CH_4), hydroxyl radicals (OH), and water (H_2O).

By now, virtually everyone is aware that the ozone layer shields the surface of the Earth from the harmful effects of direct UV radiation. However, the ozone layer is not simply an impervious brick wall. Upper-atmosphere chemistry, which one scientist described to me as "a really complicated mess," is delicately balanced and not very well understood. Solar radiation both produces and photodissociates or destroys ozone, at rates determined by the solar UV flux and the availability of other chemicals. There is reason to believe that ozone abundance varies with the sunspot cycle and the terrestrial seasons, but it is affected by so many other factors that it is all but impossible to say flatly, "This is how it works."

Nevertheless, solar radiation between 2,400 and 3,100 angstroms breaks down O_2 to produce excited free oxygen, which can recombine with O_2 to form O_3. But N_2O from the lower atmosphere is also photodissociated into N_2 and O, reducing the amount of NO and NO_2, which break down the ozone. The nitrogen dissociation is accomplished by radiation between 1,800 and 2,400 angstroms. Thus, an increase in radiation in this wavelength should increase the ozone abundance. On the other hand, an increase at the 2,400-3,100 angstroms level produces more excited free oxygen, which converts N_2O into NO and NO_2, which attack the ozone. This round-robin cycle of the creation and destruction of ozone remains fairly stable if left to its own devices.

When man enters the equation, however, things get even more complicated. Automobile exhaust, for example, heats N_2 and O_2 to produce NO, which circulates upward. In the lower atmosphere, NO reacts to create ozone; in the upper atmosphere, it destroys ozone. Supersonic transports (SSTs), it has been discovered, apparently *increase* the ozone levels, instead of reducing them, as had been feared. The SSTs operate at a low-enough altitude that the NO in their exhaust participates in ozone-creating reactions. The SST controversy arose because no one was very sure what the critical altitude was for creation versus destruction of O_3.

The reverse is true with fluorocarbons (Freon) from spray cans. These molecules are very long-lived, and when they reach the ozone layer, they apparently stay there and destroy hundreds of ozone molecules per Freon molecule. Some states have already banned Freon spray cans, but the confusion over precisely what is going on in the ozone layer has led many people to dismiss the entire problem. The makers of such spray cans take the position that the depletion of the ozone layer is not a proven fact; however, if we wait until it is proven, it will already be too late to do anything about it. As ozone decreases, more UV radiation will reach the surface, resulting in a probable increase in the rate of skin cancer, particularly at high elevations.

THE WEATHER MACHINE

The effect of all this photochemistry on the weather is uncertain. If there is any effect at all, it probably has to do with the absorption by ozone of radiation from solar flares and galactic cosmic rays. The flares and cosmic rays tend to ionize air molecules in the lower stratosphere and the troposphere. Ions serve as nuclei for condensation particles; thus, an increase in the cosmic-ray flux may lead to the creation of more high-altitude clouds. The electrical balance of the atmosphere may also be affected, which, in turn, may affect the number of thunderstorms. Or not; no one is really very sure.

We are now down to the troposphere, below 10 kilometers, where most of the weather happens. After all the filtering in the upper atmosphere, the solar radiation reaching the ground is mainly confined to the visible wavelengths between 4,000 and 7,000 angstroms. Some infrared radiation also makes it to the surface, and although we can't see it, we can feel it as heat. There is also a "radio window" at centimeter-to-meter wavelengths, which makes it possible for us to eavesdrop on the Sun and the universe through radio astronomy.

The dynamics of the weather system are even more baffling than the pyrotechnics of the upper atmosphere. Close to the Earth, solar effects are less important in creating weather than the effects of mountains, deserts, ice caps, and ocean currents. Volcanic activity can also have a pronounced effect on the weather. Traditional meteorologists take constant solar heat as a given, and go on from there. Their level of success in explaining and predicting the weather without including the Sun in their calculations is a subjective measurement. How high a rating you give them probably depends on how recently it has rained on your parade.

"The main function of the weather machine," writes Nigel Calder, "is to spread the abundant heat supplies of the tropical regions to cooler parts of the Earth." In effect, weather is the Earth's way of distributing solar energy. Equatorial regions receive more solar energy at the surface because, arriving at or near a right angle, the radiation has less atmosphere to penetrate than at higher latitudes. Heat flows away from the tropics via the complex system of prevailing winds.

The system would be much easier to understand if the Earth did not rotate. The atmosphere always lags behind the spinning globe beneath it, and Coriolis forces deflect the flow of heat from the equator, setting up intricate patterns of whirling hot and cold air masses. The speed of rotation is greatest at the equator, so the tropical air has more kinetic energy, as well as more heat. This energy is imparted to the whirling weather systems, which is one reason why hurricanes and typhoons are born in the tropics.

The warm tropical air rises, carrying with it copious amounts of moisture from the oceans. It flows generally poleward until it hits the middle latitudes, where it collides with cold polar air and begins to sink. This interaction creates the jet streams, which are relatively stable high-altitude west-to-east air flows. The jet streams prevent a direct flow of the tropical air to the polar regions; instead, the heat exchange occurs in the stormy zones of the middle latitudes. Although this is a vastly oversimplified picture of the weather system, it will serve as an adequate description of the basic elements.

Before World War II, understanding of the global weather system was meager. But the exigencies of modern warfare demanded better weather prediction, so for the first time meteorological stations were set up in remote areas. The mass of data collected provided the first good grasp of the overall dynamics of the global weather system.

Following the war, scientists realized that in order to make sense of the mass of weather data, they would have to approach the problem in a more rigorous, systematic manner. The need for more coordination was also felt in other fields of science. One attempt to achieve that coordination was the International Geophysical Year, 1957-58, during which scientists from dozens of nations participated in numerous large-scale experiments designed to provide a better idea of how the world works. Since then, there have been many similar projects, most notably the Global Atmospheric Research Program (GARP), initiated in the late 1960s. With the addition of weather satellites and computer modeling to the meteorological arsenal, weather forecasting has improved enormously.

THE MISSING SOLAR LINK

But the question of the Sun remains. Can short-term changes in the Sun somehow influence day-to-day weather patterns? The orthodox answer was (and is), "No way." An increasing number of scientists are now challenging that conclusion.

One of the leaders in the search for Sun-weather correlations was Walter Orr Roberts. A Harvard solar physicist, Roberts began exploring the problem in 1950. After participating in the IGY, Roberts went on to become the director of the National Center for Atmospheric Research in Boulder. He retired from NCAR in 1975 and is now the Director of the Aspen Institute for Humanistic Studies, which also has offices in Boulder. It was there that I spoke with his assistant and associate, Roger Olson. Like practically everyone else in the Boulder scientific community, Olson seems to exude a robust, low-key heartiness. One concludes that there must be something intrinsically healthy about the atmospheric sciences.

Olson, who got his start in meteorology at Goose Bay, Labrador, during World War II, described some of the work carried on by Roberts and his colleagues. Their search for Sun-weather links has been controversial but, by their own account, at least somewhat successful. They believe that their weather prognostications have been generally more accurate when the Sun is taken into account.

Instead of examining the entire globe, Roberts concentrated his efforts on a wedge-shaped chunk of the Pacific Ocean, above 40°N and between 120° and 180°W, near the Gulf of Alaska. For three winters in the late 1950s, the scientists looked at low-pressure areas, or troughs, passing through this region. They attempted to find a correlation between the size of the troughs and the occurrence of solar-induced magnetic storms.

Key troughs were identified as those which entered the target area three days after a magnetic storm. The normal history of a North Pacific trough is that, over the course of two weeks, it gradually gets larger as it passes over the North American continent. The Roberts group proved "to their own satisfaction" that the key troughs grew more rapidly than the non-key troughs— those not associated with magnetic storms.

The results, Olson says, "didn't create much of a splash in the scientific world." Like any statistical correlation, this one had its full share of skeptics. Even if the results were valid, their significance was open to question. Also, the correlation was with magnetic storms and not directly with the Sun. Magnetic storms were chosen as the indicator because solar flares don't always produce terrestrial effects. Flare particles travel over long, curving trajectories which are much more likely to miss the Earth than to hit it.

Roberts's results were generally ignored, so he tried again in 1970. "Walter has a tendency to do things once every ten years," says Olson. "He'll do a new project, then kind of put it away, wait ten years, and then do it again." The initial study had been criticized on the grounds that the measurements of the troughs were too subjective. But by 1970, technology and the general data base had improved to the point where it was possible to put the measurements on a more objective footing.

The scientists created a new calibration of the intensity of the troughs, called the Vorticity Area Index (VAI). They plotted lines of constant vorticity within the troughs; these were basically just measurements of the counterclockwise (cyclonic) motion of air particles. On a map, the lines of constant vorticity are roughly analogous to isobars and isotherms, which are lines of constant pressure and temperature. With computers, it was easy to measure the area within the lines of constant vorticity and come up with an objective VAI.

The new study was carried on for seven winters and "found essentially the same thing." When a magnetic storm had occurred three days prior to a trough's entry into the target area, the VAI tended to be higher. The earlier results had been confirmed. "The general scientific community," says Olson, "was maybe a little bit more interested than before, but not too much."

By this time, however, there was additional data about the Sun itself which seemed to be related to meteorological activity. It was found that the solar magnetic field extends its influence into the interplanetary medium. The field is divided into distinct sectors (usually four) in which the magnetic polarity is directed either toward or away from the Sun. As the Sun rotates, the boundaries between the sectors sweep past the Earth, causing a sudden reversal of the polarity of the field. Throughout the 1960s and 1970s, data began to accumulate which suggested a strong correlation between sector boundary crossings, and characteristic meteorological events on Earth. There was apparently a marked response in vorticity, atmospheric pressure, and electrical phenomena.

John Wilcox, a Stanford solar physicist, extended the work of Roberts and Olson and calculated the total vorticity of the atmosphere north of latitude 20°N and plotted the results against the sector boundary crossings. The findings were striking: the VAI dropped by about 10 percent on the day following such a crossing, then gradually returned to its pre-crossing level after about three and a half days.

Other studies confirmed that the effect was real. A Soviet study found that when the interplanetary magnetic field was directed toward the Sun (negative polarity), the average atmospheric pressure at a station in Siberia rose by about 5 millibars. When the field was positive, the pressure rose by about the same amount at a station in the Antarctic.

In 1974, Canadian geophysicist Colin Hines, one of the sternest critics of supposed Sun-weather interaction, undertook a reexamination of Wilcox's data. He expected to disprove the results; to his surprise, he couldn't. Olson marks this as the turning point in the search for a solar connection with the weather. Hines's international reputation was impressive, and his stamp of approval lent respectability to what had hitherto been regarded as a borderline science.

For the last five years, Olson has been concentrating on correlations between solar flares and weather activity. He has found what seems to be a link between flares crossing the central meridian of the Sun (when they are "aimed" at the Earth) and vorticity in the Earth's atmosphere. Vorticity seems to increase as the flares pass the central meridian, and then declines during the magnetic storms which follow the flares. There is also evidence that the frequency of thunderstorms follows a similar pattern.

Here, again, are intriguing statistical correlations between solar events and terrestrial weather. Scientists are now trying to identify the precise mechanisms which must be responsible for the correlations.

Although it seems doubtful that very many particles from the solar wind reach the surface of the Earth, Olson thinks it possible that a significant number of particles penetrate as far as the tropopause, the boundary between the troposphere and the stratosphere at an altitude of 10 kilometers. "This is sort of a revolutionary idea," says Olson. "Most solar particles are trapped way up in the upper atmosphere, and there's no particularly good reason to suspect that they get all the way down to the tropopause. . . . So, if you imagine that by some black magic solar particles get down to the tropopause level, then they could form condensation nuclei." The result could be an increase in cloudiness. This is difficult to confirm because natural cirrus clouds form at that altitude, and there is no immediately obvious way to distinguish between the natural cirrus and the solar cirrus.

The "ifs" get cumbersome at this point, but if solar events do increase cloudiness, the effect would be to trap more surface heat and raise the temperature locally by 1° to 1.5°C. Olson attempted to plug that additional heating into the NCAR computer model of the atmosphere. Somewhat to his surprise, the cirrus-heating had no effect at all. Olson points out that when the exercise was first carried out, in 1970, the computer models "were still pretty crude." Better models may give indications of a more subtle response to the cirrus hypothesis, but for now, "we have to put that one on the back burner."

A better candidate for the solar mechanism is the enhanced ionization of the upper atmosphere by incoming solar particles. The atmosphere and the Earth can be thought of as opposite terminals of a gigantic electrical circuit. If solar particles increase the ionization of air molecules, the conductivity of the air would also increase, resulting in more thunderstorm activity. The statistical data from flares and sector boundary crossings suggest that, over the United States, there are more thunderstorms following these solar events. Ralph Markson of MIT has made extensive studies of atmospheric conductivity as related to solar phenomena. His experimental technique is to fly his own airplane around the thunderheads and make direct measurements.

A much safer technique is to examine satellite photographs,

which is being done by Bruce Edgar of the Aerospace Corporation. Edgar counts and maps the number and distribution of lightning strokes observed from space. His tentative findings are that solar flares do increase thunderstorm activity, particularly if multiple flares are involved. The thunderstorm mechanism thus seems to be a solid candidate for the solar connection; Olson calls it "the best-bet mechanism."

The final mechanism under active consideration is—yet again —a change in the solar constant. "That," says Olson with a laugh, "would make it pretty easy."* This "brute force technique" would have obvious and immediate effects on the heat flow of the atmosphere. In reading the literature on solar-weather connections, one finds the atmospheric scientists fairly pleading for a better measurement of the solar constant. The solar scientists would be only too happy if they could comply.

Although the evidence for solar effects on the weather is mounting, it would be a mistake to conclude that the case has been locked up. The jury is still out. The cirrus mechanism, for example, may be irrelevant, even if solar particles do reach the tropopause. Some scientists believe that the availability of ions for condensation nuclei has little or no effect on cloud formation. Other evidence is being subjected to highly sophisticated statistical analysis, and the results seem to call into question some of the proposed solar-weather correlations. Steven Businger, a young scientist working with NCAR, points out that, "There is still a conspicuous lack of an acceptable physical theory linking solar activity and the weather." Businger writes that the search for such links depends on a variety of meteorological processes "about which we know so little that it is not difficult to use them to build speculative links between solar activity and weather."

EXTRATERRESTRIAL WEATHER

The difficulties involved in establishing the sought-after links become more obvious when one considers that we don't fully understand either the Sun or the weather. But there are reasons for optimism. Until recently, meteorology has been a completely earthbound science. Now, however, we are getting almost daily

* See postscript to Chapter 6.

weather reports from other planets. Rather than being limited to the confusing specifics of Sun-Earth interactions, a data base describing the general nature of Sun-planet interactions is being developed. It may still be too early to credit exo-meteorology with any major contributions to our understanding of terrestrial weather, but the exploration of other planets has at least given a new and potentially valuable perspective.

The Viking mission landed two meteorology stations on Mars in 1976. One of the Viking landers is still functioning well, and there is every reason to believe that we will continue to receive weather data from Mars for years to come. Although Martian reports are not likely to have a direct effect on tomorrow's forecast for Kansas City, the Viking observations create a kind of control experiment on the influence of oceans on the weather. Both Mars and Earth have distinct weather patterns (see page 8 of color section), but only the Earth has oceans. A comparison of data from the two planets may well give some clues about the ways in which the oceans affect terrestrial weather. It would also be important if a three-cornered correlation is found between solar events, weather on Earth, and that on Mars. Working with the Earth alone, we cannot be certain that apparent meteorological responses to the Sun are not simply unsuspected terrestrial effects. But if Martian weather is found to be sensitive to such phenomena as solar flares and sector boundary crossings, this will indicate that the work is on the right track.

Jupiter is another planet with "weather," but we are far from an understanding of the dynamics of the swirling Jovian atmosphere. The spectacular pictures from the Voyager spacecraft revealed a planet which might have been dreamed up by Salvador Dali. The differences between Jupiter and the Earth are monumental: Jupiter is huge, but it rotates rapidly (ten hours); it is far from the Sun, but it has an internal heat source; and it has a flock of giant moons which seem to exchange material with the Jovian atmosphere. But there are also similarities: the Jovian magnetic field seems to have an effect on its atmosphere; there is evidence of auroras and lightning; and it is clear that immense storm systems rage through the multichromatic atmosphere.

The Great Red Spot is the Jovian equivalent of a hurricane, several times larger than the entire Earth. (See page 7 of color section.) The Spot has existed for at least 400 years, and probably for several million years. There have been dozens of theories attempting to explain it, and even now there is no single accepted explanation. There is, however, some tentative and intriguing evidence that the size of the Spot may vary slightly with the solar cycle. The implications are interesting; if the Sun can affect storms on Jupiter, then it ought to have at least some effect on storms on the Earth, which is five times as close to the Sun.

Closer still is Venus. (See page 8 of color section.) The American Pioneer Venus and the Soviet Venera missions have revealed that the weather on Venus is hot, cloudy, and confusing. Venus has little or no magnetic field, so the solar wind interacts directly with the upper atmosphere, producing a variety of baffling effects. The Venerian weather machine seems to be much more efficient than that of Earth, spreading solar heat almost uniformly over the entire planet. Since Venus rotates very, very slowly, the atmospheric circulation mechanisms must be almost completely solar-driven.

The data from Venus, Mars, and Jupiter are yet to be digested and assembled in a way that tells us anything very meaningful about why it snows in Minneapolis in the wintertime. We have been studying extraterrestrial meteorology only for a decade or so; we have studied meteorology on Earth for centuries and still don't understand it completely. But in the coming years, the combined work of planetary, atmospheric, and solar scientists may finally give us the knowledge we need to make sense of the weather. And once we understand the weather, it may become possible to control it physically. For now, however, most people would be happy to settle for accurate and reliable forecasts. I say *most*, because weather, for all its global complexities, remains an intensely personal phenomenon. No matter what the discoveries of science, there will always be those who prefer to place their trust in the meteorological wisdom of bunions and trick knees; when it comes to the weather, you just can't please everyone.

Nine: THE SUN AND LIFE

Of all the strange and mysterious features of the Sun, the most astonishing, by far, is the fact that a miniscule fraction of its energy is responsible for the existence of living creatures on one of its planets. As far as we know, this fact makes the Sun unique in the universe.

There are good reasons for thinking that this is not such an astonishing fact, after all. As we come to a better understanding of the processes which make life possible, it seems almost inevitable that life must exist elsewhere. We have detected the raw materials of life in the depths of interstellar space and in the

turbulent atmospheres of Jupiter and Saturn. We have probed
and analyzed the soil of a nearby world and discovered chemical
processes which closely mimic the functions of terrestrial biology.
We have duplicated the environment of the young Earth and
have found in inert matter a natural preference for movement in
the direction of the essential biological reactions. What we have
not found yet, however, is concrete evidence that life exists any-
where but here.

For the moment, at least, we remain unique. Some of us would
undoubtedly like to keep things that way. If we prohibit the search
for knowledge, then our position as masters of the universe is
secure and unchallenged. We can, if we want, continue to censor
textbooks and assure their conformity with a selfish theology.
Under the guise of fiscal responsibility and the need for a balanced
budget, we can cut off basic research and take pride in our ignorance. We can draw our philosophical sustenance from the wisdom
of those famous Hollywood monster movies, whose tag line was
always the same: "There are some things that man was not meant
to know."

We can also take the opposite tack and fantasize about a universe teeming with benevolent beings who make regular visits to
Earth and impart revealed truth to its childlike inhabitants. We
can stop working so hard to understand ourselves if we believe
that all we have to do is say the right words or conjure the right
dreams, then step back and let the cosmos explain itself to us in
twenty-five words or less.

This philosophical excursion seems a necessary introduction to
the subject of the origin and evolution of life. It is a topic which
makes many people very nervous. In an age of "test tube" ba-
bies and genetic engineering, some of that uneasiness is certainly
justifiable. The acquisition of knowledge is not necessarily syn-
onymous with the acquisition of wisdom. But neither is the per-
petuation of ignorance.

At the present time a variety of chemical, biological, and geological evidence strongly suggests that life began on this planet
some four billion years ago. A belief in the validity of that evidence in no way precludes the existence of God, nor does it make
our own existence any less remarkable. The fact that certain
chemicals, given time, sunlight, and a suitable environment, can
combine to form living organisms may be interpreted as a lucky

accident, an inevitable result of natural law, or as part of a divine plan. However one chooses to view it, the creation of life is an event that is certainly worth studying.

THE ORIGIN OF LIFE

We do not know precisely when life began on this planet, but current thinking is that it arose with surprising speed following the formation of the Earth. The planet itself accreted about 4.6 billion years ago; the first living cells were probably in business by about 4 billion years ago. Allowing time for the Earth to cool and the primitive atmosphere to form, it would seem that it took no more than a few hundred million years for living organisms to evolve from inert matter.

Knowledge of the Earth before about 3.8 billion years ago is extremely limited. No rocks from that era remain on the surface. But recent evidence from the Antarctic suggests that life was present as early as 3.8 billion years ago, and we can probably assume that it predates that fossil evidence by some tens of millions of years.

Earth's original atmosphere probably consisted of hydrogen and helium, most of which was lost very early on. The Sun is thought to have gone through what is known as the T-Tauri stage, a temporary increase in luminosity accompanied by a ferocious solar wind. The T-Tauri wind cleaned up the leftover debris from the solar nebula and stripped the inner planets of their primordial atmospheres. Only the giant outer planets had enough gravity to hang onto their original gaseous envelopes.

Following the period of the T-Tauri winds, the cooling rocks of the Earth outgassed another atmosphere, rich in carbon dioxide, methane, ammonia, and water vapor. The water precipitated out over the course of millions of years, forming the oceans. During this era the Sun was presumably only about 70 percent as luminous as it is today, but the carbon dioxide in the atmosphere formed a greenhouse heat trap, maintaining temperatures above the freezing point of water.

Early in the twentieth century, scientists began to wonder how this early inventory of gases could have combined to form living organisms. Darwin had postulated the existence of some "warm little pond" teeming with complex organic chemicals necessary for life. Over eons, the organics would form millions of random combinations, sooner or later hitting on the precise mixture necessary for life. In the 1920s the Russian scientist A. I. Oparin expanded on this concept, putting forth a theory of "chemical evolution."

But the random workings of chance seem insufficient to account for the development of complex proteins and nucleic acids. If you put all the parts of a Mercedes into a huge bag and shake it, it is unlikely that mere chance will ever produce a functioning automobile.

In the early 1950s, American chemist Stanley L. Miller performed a simple experiment which provided the glimmerings of an answer to the problems posed by random chemical evolution. He mixed together a combination of water vapor, methane, ammonia, and hydrogen—a "model" of the early atmosphere—and exposed it to an electric spark for a week. The process resulted in the formation of a brown sludge which contained high concentrations of the amino acids glycine and alanine, which are basic components of most proteins. Variations on this experiment have been performed many times, producing a broad inventory of essential amino acids. It seems that, given the right conditions and plenty of time, the compounds available in the early atmosphere preferentially formed the precursors of life. This discovery made the formation of life seem much less of a random shot in the dark.

The Miller experiment did not explain everything. There were still major problems having to do with concentrations, the rate of formation versus the rate of destruction, and the precise mechanisms by which the amino acids could give rise to the necessary nucleic acids. Many of those questions are still to be answered. We need not get into the baroque details of biochemistry here; it is sufficient to say that by four billion years ago, the stage was set for the appearance of life.

The role of the Sun is difficult to determine. With no free oxygen in the early atmosphere, a protective ozone layer could not have formed. Ultraviolet radiation must have bathed the surface of the young Earth. Organic molecules tend to be destroyed by UV. The prebiological reactions must therefore have taken place either in deep water or possibly in the lattices of sedimentary clays—this point is in some dispute.

If solar radiation destroyed organics, it may also have provided the initial energy needed to promote the creation of the first amino acids. However, the Sun was not the only available source of energy. Lightning could have done the job; we know that electrical discharges are common in the atmospheres of Venus and Jupiter. Another possibility is that the source was the acoustic energy of thunderclaps produced by the lightning.

Not all scientists are convinced that life first formed in the organic soup of the early oceans. There is some support for the "panspermia hypothesis," which suggests that the original source of organic material may have been meteorites, comets, or even the waste products of visiting astronauts. We do know that organic material is present in some meteorites, although nothing resembling life has been found in extraterrestrial material. This is an intriguing hypothesis, and not at all impossible, but it doesn't really seem necessary.

THE PRECAMBRIAN ENERGY CRISIS

Whatever the original source, and whether the essential ingredients formed just once or many times, it seems clear that before life itself was present, there was a seething chemical ecology at work. Certain complex molecules would have an evolutionary advantage in the early environment. If molecule A reacted with competing molecules B and C to produce more of molecule A, it would quickly come to dominate the scene. A prebiological natural selection process was at work, weeding out the less efficient combinations.

It is difficult to define an unambiguous threshold separating life from nonlife. During the Viking mission to Mars, the biology experimenters were faced with this very problem: how do you recognize life when you see it? They discovered in the Martian soil some extremely complex chemistry, which in many ways resembled terrestrial biochemistry. After months of debate and laboratory experimentation, they concluded that what they had observed on Mars was chemical rather than biological, but the dividing line was somewhat fuzzy, and some scientists think the question is still open. A time-machine mission to the early Earth might produce similar ambiguities.

The first living organisms were prokaryotic heterotrophs. That is, the early cells had no nuclei (prokaryotic) and were unable to manufacture their own food (heterotrophic). They survived by consuming the organic molecules in the primitive sea and extracting their chemical energy through the process of fermentation. This is a highly inefficient way of getting energy, but it served the purpose. The organics were broken down into carbon dioxide and alcohol, which were excreted from the cells. As long as there was an abundant supply of organics, the early cells prospered.

At this point, the plot takes a dramatic and instructive turn. The supply of organics was limited. As the population of heterotrophs increased, the food supply diminished. Our earliest ancestors were creating their own energy crisis. As their "fossil fuels" disappeared, they were faced with extinction. Our unicellular forebears survived only by learning to utilize solar energy.

PHOTOSYNTHESIS

Between three and four billion years ago, the crisis became acute. But somehow, somewhere, photosynthesis was invented. Cells developed which were capable of manufacturing their own food by internally combining carbon dioxide, water, and the energy from sunlight. The fundamental equation is:

$$6 \ CO_2 + 6 \ H_2O + \text{sunlight} \rightarrow C_6H_{12}O_6 + 6 \ O_2$$

The sugar is then broken down to extract chemical energy, and the oxygen is excreted.

Not only was photosynthesis a way out of the energy crisis, it was also far more efficient than fermentation. Organisms capable of photosynthesis thus had two distinct evolutionary advantages over organisms which relied on fermentation. They had a steady and unlimited source of energy and food, and they had more internal energy available for such important activities as locomotion, growth, and reproduction.

There was still another important advantage to photosynthesis: its waste product killed off the competition. Oxygen is a deadly poison for anaerobic (fermenting) organisms. As the population of photosynthesizers increased, so did the oxygen content of the ocean and the atmosphere. The rate of oxygen production is a controversial subject, but it seems likely that 3 billion years ago the level of atmospheric oxygen was less than 0.001 of the present abundance; by 1 billion years ago, the O_2 level had probably increased to about 0.1 of the present amount—a hundred-fold jump.

Times were tough for the nonphotosynthesizers. Most of them probably became extinct during this period. But, in another useful lesson from Precambrian history, some of the early organisms discovered that cooperation can pay better dividends than competition. A symbiotic relationship arose between anaerobic microbes and the photosynthesizers. The anaerobes somehow began to incorporate the aerobic organisms into their cells, an arrangement that was advantageous for both. The photosynthesizers found a safe home and the anaerobes acquired a means of dealing with the poisonous oxygen.

From this humble beginning, the prokaryotes eventually evolved into eukaryotes, or nucleated cells. With the appearance of a structured nucleus, mitosis became possible. Mitotic cellular division made reproduction a more formal, organized activity, accelerating the process of evolution through random mutation of the nucleic acids in the nucleus. It took nearly 3 billion years to get to this point; in the 1 billion years since then, we have gone from one-celled algae to multicellular plants and animals, and such complex and highly evolved organisms as trees, flowers, insects, reptiles, birds, monkeys, and human beings.

Today, we see the remnants of that early symbiosis in the chloroplasts and the mitochondria. These tiny structures within nucleated cells retain a DNA structure different from that of the cell itself. The chloroplasts and mitochondria are, in effect, alien organisms which have colonized all eukaryotic cells.

Photosynthesis is carried on in the chloroplasts. The vital ingredient in the chloroplasts is the chlorophyll molecule, which absorbs solar radiation in red and violet wavelengths. It reflects the intermediate wavelengths, which is why most plants are green. The fact that photosynthesis is powered by the most abundant wavelengths of solar radiation is indirect evidence that the Sun has never undergone radical changes. Photosynthesis probably would not have evolved in its present form if the Sun had not been a dependable source of radiation in the visible wavelengths.

We are still attempting to understand the details of photosynthesis. Basically, photosynthesis is a process which provides the energy necessary for cells to manufacture molecules of adenosine triphosphate (ATP), which then store the energy for future use within the cells. The chloroplasts chemically split water mole-

cules, making free electrons and protons available from the hydrogen. The energy of incoming solar photons drives the electrons
across a series of highly structured membranes within the chloroplasts via a series of "carrier" molecules. At the end of the process,
the electrons and protons recombine with carbon-dioxide molecules to form organic molecules and ATP. The ATP then releases
energy by detaching a phosphate group and becoming adenosine
diphosphate (ADP).

The precise roles played by the chloroplast membranes and the
carrier molecules are still under investigation. Apparently two
photons must be absorbed in order to drive each electron across
the membrane. The absorption is accomplished by a specialized
chlorophyll molecule known as P-680, which absorbs radiation at
6,800 angstroms. At an intermediate stage in the reaction, more
photons are absorbed by another chlorophyll molecule, P-700, and
the reaction proceeds through additional carrier molecules to the
ultimate synthesis of ATP. All of this happens very quickly, within
about 1/100,000,000 second.

With the invention of photosynthesis, biology became an important factor in the transformation of the young Earth. Photo-
synthetically produced oxygen accumulated in the atmosphere,
creating an ozone layer. With a shield against ultraviolet radiation, life was able to move into shallow water and onto the land.
The rate of oxygen production increased. During this period the
faint early Sun was gradually warming up, increasing the amount
of solar radiation available for photosynthesis.

At this point, we can again invoke the Gaia Hypothesis (see
Chapter 7). The rates of oxygen production and increasing solar
luminosity fit each other almost too well. If too much oxygen was
created too early, the greenhouse effect would have been lost and
the global temperature would have plunged. If oxygen production
had proceeded too slowly, the greenhouse would have trapped
too much heat. It would seem that some biological feedback mechanism must have modulated the atmospheric oxygen content in
order to keep the temperature suitable for life.

DEATH OF THE DINOSAURS

The necessary feedback mechanism is probably extinction. The
geological record of the last billion years shows that there have
been periodic waves of extinction which decimated the population

of living species. If the photosynthesizers were doing too good a job of producing oxygen, the climate would change and kill off enough organisms to get things back in balance. This theory is supported by some recent evidence which indicates that there have been major fluctuations in the oxygen content of the atmosphere.

From a geological viewpoint, these biological adjustments were sudden and catastrophic. However, if you were living in the middle of such an episode, you probably wouldn't even notice it. According to atmospheric scientist George Reid of NOAA, if you killed all the plants in the world tomorrow, oxygen would remain in the atmosphere for some 2 million to 3 million years.

We should be careful, however, not to ascribe too much importance to this biological trigger mechanism. There are many other factors which can account for the widespread extinction of species. The drift of the continents was undoubtedly responsible for the extinction of many species which lived in the shallow waters bordering the continents; as the continents crashed together, the total area of shallow offshore waters decreased. Continent building was also accompanied by areas of intense volcanism, which had a variety of effects on the atmosphere. Finally, the Sun and the stars may have played a major role in the extinction of some species.

As organisms emerged from the ocean, they exposed themselves to higher levels of radiation. Even the ozone layer does not remove all of the solar UV output. In addition to the ultraviolet light, organisms were also exposed to particles from intense solar flares and to galactic cosmic rays. From an evolutionary standpoint, all of this is good. Radiation creates genetic mutations, and although the vast majority of such mutations are harmful, a few of them produce biological improvements. The rate of evolution accelerated as life spread over the land.

Among the species which evolved were the dinosaurs. We often think of dinosaurs as stupid, slow-moving brutes which were incapable of adapting to a changing environment. But the evidence suggests otherwise. The dinosaurs flourished for over 100 million years, which, by some standards, makes them a much more successful species than *Homo sapiens*. But 65 million years ago, abruptly and mysteriously, the dinosaurs vanished. Paleontologists have been attempting for over a century to explain this mass extinction, and although a variety of ingenious explanations have been proposed, we still are not sure what happened.

There is a growing body of evidence that the dinosaurs may

have been destroyed by a sudden, fatal increase in radiation. This could have been accomplished in a number of ways. As mentioned earlier, the Earth's magnetic field periodically reverses its polarity. We know this reversal has happened many times throughout history, and we can even tell when it happened. As rocks cool and crystallize, their magnetic orientation is frozen according to the prevailing polarity and the position of the magnetic poles. By examining the geomagnetic record in various rocks, we can determine the polarity during the period when the rocks crystallized; this is also a method of charting the motions of the continents. When a polarity reversal occurs, the general magnetic field of the Earth becomes extremely weak. For a period of a few thousand years, the magnetic field is insufficient to protect the surface from incoming solar and galactic particles. It has been estimated that the increased radiation flux would multiply atmospheric ionization by a factor of 10. The result of the intense ionization would be a plunge in the level of ozone. Thus, during a magnetic reversal, organisms dwelling on land or in shallow water would be exceptionally vulnerable to radiation.

This is a controversial point, but recent work by J. D. Hays has confirmed that there is a definite statistical correspondence between recent polarity reversals and the frequency of extinctions. Only a few thousand years of increased radiation might be enough to wipe out a species which already had a marginal existence. But the dinosaurs were well established; it seems unlikely that they could be eliminated so quickly.

The dinosaurs, however, were not the only ones to disappear. At the end of the Cretaceous period, about 65 million years ago, approximately one-third of all living species suddenly became extinct. Clearly, something extraordinary happened.

There are two good candidates for the killer mechanism: solar flares and supernovas. Geological evidence indicates that preceding the wave of extinctions, there was a long period during which there were no polarity reversals. At the end of the Cretaceous, the reversals suddenly resumed. If these reversals coincided with an episode of high solar flare activity, or the occurrence of a nearby supernova explosion, the background cosmic radiation level could have jumped a thousandfold.

A supernova within 10 parsecs of the Earth would have been particularly devastating. The best estimate is that such an event can be expected to occur about once every 100 million years, and there is now reason to believe that the last one happened at the end of the Cretaceous. American nuclear physicist Luis W. Alvarez, and his son, Walter, a paleontologist, have analyzed late Cretaceous rocks and found an anomalously high abundance of the rare element iridium. Iridium is one of only four elements which have a significantly higher abundance elsewhere in the universe than on Earth. There is about twenty-five times less iridium in terrestrial rocks than the normal cosmic abundance. Yet the late Cretaceous rocks show that for a brief period, terrestrial levels of iridium equaled the cosmic levels. One possible source for the introduction of the anomalous iridium would be a nearby supernova.*

George Reid and his co-workers at NOAA and NCAR have studied the possible effects of a temporary thousandfold increase in radiation. One major effect would be a depletion of ozone and an increase in the level of stratospheric NO_2. This change in the chemistry of the upper atmosphere would alter the spectral absorption properties of the atmosphere, resulting in "an appreciable decrease in the total integrated solar radiation available for heating the lower atmosphere and the surface." The effect would be most pronounced in the blue region of the spectrum, which is where plants get much of their energy for photosynthesis.

Adding it all up, we get an impressive picture of global catastrophe. Direct radiation reaches the surface in lethal doses. The climate gets colder and probably drier. Photosynthetic oxygen production declines. This would be very bad news for anything living on the Earth, but especially for dinosaurs. Despite some recent suggestions to the contrary, most—if not all—dinosaurs were cold-

---

* The supernova and flare theories are being challenged by a mass of new evidence. The Alvarezes and other investigators now believe that the iridium may have been deposited by a global dust cloud produced by the impact of a small asteroid, perhaps 10 kilometers in diameter. Such a collision would have had devastating effects on terrestrial life and could possibly account for the extinction of the dinosaurs and other organisms. The collision theory, however, works equally well with or without terrestrial polarity reversals. The timing of the reversals is intriguing, and it is probably too soon to rule out a lethal increase in solar or galactic radiation.

blooded. They would not have thrived in a colder climate. Lower oxygen levels would also have been particularly harmful to them. We can, at this point, even bring in some of the other theories explaining the demise of the thunder lizards. Mammals had been around for quite some time, but they were still small and probably didn't provide much competition. But the early mammals were almost certainly nocturnal. If solar flares were responsible for the increased radiation, the mammals would have been relatively safe in their burrows during the dangerous daytime. Coming out at night, they could feed on the dinosaur eggs and generally make life even more unpleasant for the doomed behemoths.

Much of this theory is still conjectural, but it does seem to provide a tidy explanation for why the dinosaurs and so many other species died out so suddenly. A supernova explosion would seem to be the most likely source of the deadly radiation, but the Sun cannot be eliminated. Given the neutrino problem, we can't say with any certainty that the Sun has not undergone periods of violent flare activity. We do know that it has experienced extended periods of very low activity, during which the solar shield against cosmic radiation was less effective. If an era of high flare activity coincided with a terrestrial polarity reversal, or if a period of very low activity coincided with a nearby supernova, the stage would be set for the dinosaurs' final act.

SOLAR ECOLOGY

The Sun may have become a killer at times, but for at least the last 65 million years it has been a benign and dependable source of energy for living organisms. Virtually all of the energy used by the biosystem begins in the Sun. There are a few remaining species of anaerobic organisms, but even they are ultimately dependent on the Sun for their supply of dead organic matter. Recently a handful of maverick species have been discovered living in completely nonsolar environments, but they are rare indeed. Some microbes live in the sulfurous waters of geysers and hot springs; others have been discovered in the waters off Baja, California, and the Galapagos Islands, living off the energy of volcanic vents. Undoubtedly, there are other organisms yet to be discovered, deriving energy from unexpected sources. It is probably not impossible that some organisms may get their energy from naturally occurring radiation sources. But for the vast majority of organisms, the Sun is the alpha and omega of existence.

In the last few years, scientists have begun to study the ways in which solar energy is distributed throughout the biosystem. The traditional "cats eat rats" approach to ecology is giving way to sophisticated techniques of systems analysis. Ecologists now divide their time between the wilderness and the computer terminal, where they produce complex energy distribution analyses which resemble paper flow charts in the Pentagon. It may seem incongruous to reduce the great outdoors to lifeless data bits, but the new methods may ultimately do more to preserve the environment than all the well-intentioned efforts of birdwatchers and nature lovers.

The computers are necessary because there are more variable factors to consider in ecology than in any other natural science. A researcher may spend years making observations in the wild, but he cannot take a forest or a marsh back to the laboratory with him. What he can do is use his observations to make a detailed mathematical "model" of the ecosystem. Once the validity of the model has been verified by comparisons with nature, it becomes possible to isolate the various elements of the ecosystem artificially and analyze the roles they play. Changing parameters can be introduced into the model to study the effects of variable factors such as weather, pollution, agriculture, hunting, logging, forest fires, or even (in some frightening scenarios) nuclear war. The models are not one hundred percent accurate, but they do provide a limited predictive capability, allowing mankind to plan for the consequences of environmental changes before it is too late to do anything about them.

The quantity of solar radiation reaching the biosphere is the fundamental limiting factor on the activities in an ecological system. The total amount of insolation varies according to a number of factors, including latitude, air quality, weather conditions, and the behavior of the Sun itself. The distribution of the solar energy within the ecosystem varies in relation to the types of vegetation present. Some plants are good photosynthesizers, while others are less efficient. In order for the solar energy to be made available for use by other inhabitants of the ecosystem, the plants must first convert the radiation to carbohydrates. In the wild, the overall efficiency of conversion seems to be between 1 and 2 percent. The same efficiency is typical in agriculture, although some scientists believe that an efficiency of 8 to 10 percent is theoretically possible. If we can discover ways to maximize photosynthetic efficiency, food production could be increased substantially.

The solar energy fixed by green plants may enter the rest of the ecosystem in a number of ways. Animal grazing may account for most of the immediate distribution, but the amount of energy involved may vary dramatically from year to year, depending on the local population of insects and herbivores. For example, a recent study of the Hubbard Brook Experimental Forest in New Hampshire found that in most years, only 1 percent of the fixed solar energy was distributed through grazing; but when the population of a certain species of caterpillar was high, the figure could reach more than 40 percent. Thus, variable environmental factors, such as caterpillars, may have a profound and immediate effect on energy distribution.

Most of the fixed solar energy is stored, either in the living biomass or in the dead organic matter that accumulates in the soil. In the Hubbard Brook study, it was found that the organic debris contained about 1.7 times more energy than the living plants. A variety of bacteria, insects, and small animals feed on these dead organics in the soil. In a healthy ecosystem, slightly more energy is stored than eaten, resulting in a gradual accumulation of energy in the soil. This reserve is an important factor in the continued survival of the ecosystem. Ultimately, the most destructive effect of fires and logging operations is that they expose this energy reservoir to increased erosion, removing vital energy stores from the environment.

The energy that is not consumed or removed will remain in the soil. If it remains there long enough, it may eventually become coal, oil, or natural gas: permanently fixed forms of solar energy. When we burn a gallon of gasoline, we are simply recovering a minute percentage of the energy of solar radiation that arrived here a hundred million years ago. This is, in many ways, convenient. It is also, in many ways, idiotic.

The Sun, after all, is still shining.

Ten: SOLAR SOLUTIONS

After all our worries about neutrinos and flares and oscillations and UV variability, it remains a fact that in two weeks, we receive more energy from the Sun than is stored in all the fossil fuel reserves on the planet. And we waste almost all of it.

Solar energy is dependable, inexhaustible, and essentially free. Our access to the Sun is not controlled by the whims of the Ayatollah Khomeini, the Sultan of Oman, or the Sheik of Araby. Exxon does not yet own the Sun. No government bureaucracy parcels out solar energy allocations. There is no known tax on the Sun. Solar energy does not pollute the air or the water. There are no defective valves in the Sun. Solar waste will not give your great-grandchildren cancer.

The Sun is mankind's oldest source of energy. Before the Industrial Revolution, it was virtually our only source of energy. Today, however, even the indirect uses of solar energy provide only a small percentage of total power consumption. Less than 5 percent of United States energy needs are met by direct and indirect solar sources; worldwide, the figure is probably between 15 and 20 percent. The rest of our energy, aside from some very limited use of geothermal sources, comes from the combustion of fossil fuels and the fission of uranium atoms.

The fossil fuels will be gone soon. There is considerable debate about how soon "soon" will be, but from the long-term perspective, it is immaterial whether the oil wells dry up in twenty years or two hundred. Coal reserves may stretch another four hundred years, but that, too, is irrelevant. We are rapidly approaching the crisis faced by our anaerobic ancestors some three billion years ago; they learned to use other sources of energy, and we must do the same.

Our choices are complex and bewildering, but in the end, it seems that we have only two ways to go. We can use the Sun, or we can mimic it. The art of mimickry is not easily learned, nor is it foolproof. By splitting or crushing atoms, we are attempting to do on a small scale what the Sun does on a cosmic scale. The achievements of nuclear technology are monumental and, in some ways, inspiring. Creating and controlling dozens of our own mini-Suns is no small accomplishment. But why buy a copy when we can use the original?

In recent years we have rediscovered the Sun. Interest in solar energy has risen at least as fast as the price of oil. In the process, a new solar mythology has also evolved. Out of optimism, ignorance, or paranoia, increasing numbers of people are coming to think of the Sun as a gigantic electrical socket in the sky. It's the Big Fix we've been waiting for. Limitless energy, too cheap to meter, there for the taking—safe, efficient, and clean.

Twenty-five years ago, people were saying the same things about nuclear energy.

We have learned—after the fact—that nuclear energy is not exactly what we had been promised. It would be unfortunate if we were to make the same mistakes in our attitudes about solar energy. The realities of solar energy are not quite as glittering as

our dreams; nor are they as depressing as the oil companies and
utilities would have us believe. There is no single "solar solution"
to our energy problems, but there are many different paths open.
Some of them may lead to the solar future of our dreams and
hopes; others may be dead ends. Whatever choices we make, getting from where we are to where we want to be will not be easy.

The promise of solar energy lies in its diversity. It can be used in
many different forms, for virtually any purpose we can imagine.
Hydroelectric power is solar energy once removed. So is the
methane derived from garbage, and the electricity from photovoltaic cells. The successful transition to solar energy will depend
not on brute force methods of squeezing every last watt out of
the Sun, but on our willingness to utilize appropriate solar technologies when and where they are needed.

Because the concept of solar energy embraces so many different
techniques of energy production, it is a decentralized technology.
Useful research is being carried on by major corporations and universities, but also by shoestring enterprises and backyard engineers.
Increasingly, the United States government is becoming involved
in solar research—a mixed blessing. Federal dollars support many
promising lines of research, but the viability of a particular project
may ultimately come to depend on the judgments of bureaucrats
rather than those of scientists and engineers.

SERI

The major responsibility for coordinating American solar research
lies with the Solar Energy Research Institute (SERI) in Golden,
Colorado, just west of Denver. SERI was born in the confused
aftermath of the 1973 Arab oil embargo. In 1974 Congress passed
the Solar Energy Research, Development, and Demonstration
Act, a provision of which called for the establishment of a national
solar energy research center. The Energy Research and Development Administration (ERDA) called for proposals for a "Government Owned Company Operated" (GOCO) facility. The
winning proposal was submitted by the Midwest Research Institute, a Kansas City "think tank," and the contract was let in
March of 1977. SERI began operations in July of that year, under
the direction of Dr. Paul Rappaport. SERI now receives funding
from DOE.

When I visited SERI in the summer of 1979, the center employed some 600 people, who were chosen from over 12,000 applicants. A high percentage of the SERI staff consisted of young people, in their twenties and thirties, all of whom seemed enthusiastic about the prospects for solar energy. Not long after my visit, Dr. Rappaport stepped down as head of SERI and was replaced by Denis Hayes, a "grass roots" solar activist. To many people in the field, the appointment of Hayes was seen as a significant commitment to the development of practical solar technologies. But regardless of who runs SERI, its budget is still controlled by DOE, which makes it subject to all the political, economic, and bureaucratic entanglements that are inherent in government operations.

Within SERI, there are "600 different views" about what solar policy should be. The Carter administration has stated—with a notable lack of consistency—that it should be possible to provide 20 percent of our energy needs through solar sources by the turn of the century. Depending on who is speaking, that 20 percent figure can shrink substantially. The consensus at SERI seemed to be that 20 percent is a realistic and achievable goal.

How we go about reaching that goal is still a subject of considerable debate. The diversity that makes solar energy so promising also makes it confusing. For policy planning, the DOE considers solar energy to consist of any energy source derived from "recently arrived" sunlight—recently meaning within the last 100 years. There are three broad areas of study: solar heating and cooling, fuels from biomass conversion, and solar-generated electricity. Differing types and levels of technology are involved in each area, making comparisons of cost and efficiency difficult. With so many different directions to go, plotting a precise course is all but impossible.

SOLAR HEATING AND COOLING

Solar heating and cooling is the most immediately productive field. The technology involved is, for the most part, uncomplicated and inexpensive. The Romans heated their public baths with solar energy. The Greeks planned their cities to take advantage of solar heat, as did the Mesa Verde Indians. These early civilizations appreciated a fundamental fact that seems to have escaped modern architects and builders—to get heat from the Sun, all you have to do is face it.

"People don't even understand the benefit of orienting a building to face south," says Bruce Baccei of SERI's Passive Technology Branch. In this case, "going solar" is simply a matter of basic education. "The goal," says Baccei, "is that every person in the United States who understands 'one and one is two' and 'ABC'— just that, not even the whole alphabet—ought to understand the advantages of a south-facing building."

A building with the long axis oriented east-west reaps a double benefit. In the winter (in the Northern Hemisphere) the Sun stays low in the southern sky, and provides direct heating to the south-facing windows. In the summer, the Sun is higher in the sky and its major impact is on the roof and the short east and west sides of the building, where there should be few windows. The south-facing house thus gets heat from the Sun in the winter and shade from it in the summer. Simple but effective.

The advantages of east-west orientation apply in every climate. "It's reasonable," says Baccei, "that you can provide from fifty to eighty percent solar heating in virtually every climate in the United States." Even in areas such as New England and the Pacific Northwest, the Sun shines for more than 2,000 hours each year. In the Southwest, annual sunshine averages from 3,000 to 4,000 hours. These facts are fundamental, but they have had remarkably little impact on builders, architects, and the general public, who have literally been turning their backs on the Sun.

Obviously, we are stuck with a lot of buildings that don't face south. But virtually any building can be "retrofitted" to take advantage of whatever Sun it does receive. This does involve the application of technology, but not of the twenty-first-century variety. Even nineteenth-century technology will do. At the turn of the twentieth century, before the advent of cheap, plentiful oil, thousands of houses in the Los Angeles area utilized rooftop solar water heaters. Today we are rediscovering the benefits of simple solar technology. The solar heating industry in the United States has grown enormously in the last decade and is now turning over nearly half a billion dollars annually.

The basic solar system is a flat plate collector, usually installed on the roof. The collectors may consist of metal sheets (copper, iron, or aluminum) which are coated with a substance such as carbon black to make them as dark as possible in order to absorb 95 percent or more of the solar radiation. The plate transfers its heat to fluids circulating behind it.

This photograph shows a house in Washington, D.C. that was "retro-fitted" with solar panels for a solar space heating system. The total cost of the conversion was $2,500; the owner estimates that he has reduced his dependence on oil by two-thirds, and that if his house had been more favorably sited, the costs would have been half as much.

U.S. Department of Energy

In the most basic system—the thermosiphon—the fluid is water. It flows downward from a tank into the collector, where it is heated and flows back up into the top of the tank. The thermosiphon system is completely automatic and requires no controls or technical wizardry. Its drawback is that it won't work in areas where freezing temperatures occur. Although this effectively excludes most of the United States, basic thermosiphon systems have been used successfully for years in other parts of the world.

In colder areas, the system requires a pump and a few controls. In a basic "drain down" system, an electronic sensor activates a drain valve when the temperature approaches freezing. This removes the water from the system, protecting the collectors against the freeze. To get around the temperature problem, most new solar water heaters employ an antifreeze solution in the collectors. The antifreeze then flows through a heat exchanger to pass the heat along to the water.

An alternative to the flat plate collector is the concentrating or focusing collector. This system employs a parabolic reflector which focuses solar radiation on a small section of piping. Concentrators can achieve much higher temperatures than flat plates, but they are less effective on cloudy days. Flat plates can absorb scattered solar radiation and don't require direct sunshine.

"The wave of the future," according to SERI's Arnon Levary, seems to be the evacuated tube collector. This system resembles the concentrator, but the sunlight is focused on a vacuum tube, rather than directly onto a pipe. The water pipe runs through the vacuum tube and absorbs the heat, while the insulation provided by the vacuum cuts down heat loss. Evacuated tube collectors are already being produced commercially, but they are not likely to become popular until the cost can be reduced.

In the more distant future, we may see a switch to "black solar fluid" systems. In basic systems, the water is two steps removed from the direct solar heat; first the collector plate is heated, then the anti-freeze, and finally the water. The black solar fluid, if perfected, would remove the first step. Instead of getting secondhand heat from the collector plate, the black solar fluid would be exposed to the Sun directly. The problem here is to develop a fluid which will efficiently absorb the direct solar radiation. This is one area in which a backyard tinkerer may be as likely to achieve success as a big corporation. Organized research is being conducted, but a lucky amateur might well be the first to hit on the best combination of ingredients for an effective black solar fluid.

Solar energy can also be used for space heating by means of a system similar to that for hot water. Some space heating systems concentrate solar radiation on air pipes which distribute the heat throughout the building. Air systems have the advantage of being freeze-proof, but they tend to be bulky and noisy. Hot water can be produced as a by-product of the air systems, but the maximum temperature is only about 140°F. Solar space heating is not yet economically attractive because there are no "package" systems available. Each system must be custom-designed for the building in which it is to be installed.

In both water and space heating a critical factor is storage. There are peaks and valleys inherent in virtually any solar energy system. Adequate provisions must be made for the times when the Sun is not shining. However, in terms of economics, it pays not to build too big a storage tank. Twelve hours of storage capability seems to be the practical maximum.

For most parts of the United States, conversion to solar heating systems will not mean complete emancipation from traditional energy sources. "Any solar system you have," say Levary, "whether it's hot water or space heating, still needs backup." Depending on the location, a solar system may save from 50 to 80 percent on conventional heating bills. With the rising price of oil, even a 50 percent saving is economically appealing. A study by the Harvard Business School estimates that by the turn of the century, active and passive solar heating systems could provide the equivalent energy of 3 million barrels of oil per day. At the current price of oil, that amounts to a national saving of $100 million a day.

Most people, however, are more concerned with their own energy bills than with the nation's. Installing a solar heating system is not cheap, and the pay-back period is uncertain. On the average, Americans tend to move every five years; thus, to be attractive, a solar system ought to have a pay-back period of less than five years.

A simple hot-water system can be installed for a few hundred dollars. More elaborate systems are in the $2,000 to $3,000 range. Combination space and water heating systems may cost more than $10,000. Calculating the pay-back period for any system is difficult because so many factors are involved—from local tax rates and personal income tax bracket to the prime lending rate and the world price of oil. Identical systems in different locations may have pay-back periods that differ by a factor of two. Although the federal government and many states have offered tax breaks for solar users, the lack of a consistent, overall policy compounds the general confusion. In some areas, for example, utility companies are permitted to charge solar owners higher rates for the use of conventional back-up power.

There is no doubt that the general public thinks of solar energy as an attractive alternative, but at the present time the uncertainties are so great that many people are holding back. Some people are undoubtedly waiting for the "calculator effect" to take hold. When pocket calculators were first introduced, they cost hundreds of dollars; a few years later they were selling in supermarkets for ten dollars. Unfortunately, nothing of this sort is likely to happen in the solar marketplace. Costs aside, the basic technology already exists, and has existed for years, and no fundamental price-shattering breakthroughs are on the horizon.

As SERI's Jerome Williams explained it to me, the life cycle of any product involves distinct phases. The first to buy are the innovators—the gadget freaks and tinkerers. Then come the early adopters, the late adopters, and the laggards. At the moment, solar energy is still in the early phases. The first person on your block to go solar will probably be the same guy who had the neighborhood's first color television or home computer. The innovators tend to buy on the basis of noneconomic factors. To reach the mass market, however, the monetary advantages of solar energy will have to be more readily apparent.

"People's concept of what solar is all about is very limited," says Williams. "People generally equate solar with putting a collector on your roof, and that's a very small part of what the potential of solar really is." Still, the rooftop collector is the most accessible facet of solar energy, and the public's acceptance of it will be crucial in determining the success of broader solar applications. The general confusion about home solar systems thus has a ripple effect on the entire field of solar energy. Clearing up that confusion is one of SERI's major goals, and it would seem that progress is being made. At the Government Printing Office bookstore in Boston, for example, the best selling books and pamphlets are those which deal with solar energy. Steadily, and no longer slowly, the public is becoming more aware of the solar potential.

Industry is also beginning to turn toward the Sun. Industrial users consume about 40 percent of the nation's energy. A SERI study estimates that between 50 and 70 percent of the industrial use is in the form of Industrial Process Heat (IPH), which is defined as "thermal energy used in the preparation and treatment of goods produced by manufacturing processes." Existing solar systems are most effective at producing temperatures lower than 550°F. Industrial processes requiring temperatures in that range account for about 27 percent of industrial energy use. If solar preheating for higher temperatures is included in the calculation, solar energy could provide as much as 51 percent of industrial energy needs.

It seems unlikely that American industry will undertake a massive conversion to solar energy in the near future. There are many technical and economic factors working against the large-scale use of solar energy in industry, not the least of which being the huge capital investment in traditional energy systems. Although some major corporations, such as Boeing and Grumman, have made significant commitments to solar technologies, most of the entrenched heavyweights of American industry have a vested interest in *not* going solar.

Despite the obstacles, there are already a number of commercial operations employing solar energy. A Campbell's soup plant in Sacramento, Calif., uses solar heat to wash cans. The York Building Products Co. in Harrisburg, Pa., uses a multiple solar reflector in the curing of concrete blocks. These and several other industrial projects are partially funded by the DOE, but a few privately funded solar operations also exist. The Budweiser brewery in Jacksonville, Florida, employs an evacuated tube collector in pasteurizing its beer. Many smaller firms are also making limited use of solar energy.

To this point, we have been talking mainly about the popular rooftop collector concept of solar energy. These systems are already available and can make an important contribution to the solution of our energy problems almost immediately. For the near future, however, solar energy may make its most significant advances in the field of biomass conversion.

BIOMASS CONVERSION

The biomass, in essence, consists of anything in our environment produced by living organisms. The use of biomass is, of course, nothing new. Woodburning stoves are the prime example of our historical use of solar-biomass energy. But trees take a long time to grow and it's not necessarily a good idea to use them as a source of heat. Better ways of getting energy from plants are being developed, and the potential of biomass conversion is enormous.

The simplest way to get energy from the biomass is to burn it. Economically, however, burning is probably the least productive use of biomass material. If I chop down a tree and burn it, I get less value from the tree than if I had sawed it into lumber or sold it to a paper mill. A much more effective use of the woody biomass is achieved by chemically converting it to methanol or methane gas.

Biomass conversion to liquid or gaseous fuels will be a key

factor in the transition to solar energy. About one quarter of our energy is used for transportation, and most systems of transportation require liquid fuels. A gradual changeover from fossil to solar liquid fuels could be accomplished with relative ease; indeed, it is already happening in many areas. Similarly, we already make substantial use of natural gas, so a large-scale shift to solar methane would require no fundamental restructuring of our energy distribution system.

Producing alcohol from grains is one of civilization's oldest activities. Using the alcohol for fuel instead of libation is also nothing new. During World War II, the Germans made extensive use of alcohol from potatoes to run automobiles and other vehicles. Internal-combustion engines can run effectively on gasoline-alcohol blends without major modifications. Gasohol is already available at the pump in most parts of the country. It is presently more expensive than straight gasoline, but the initial reaction to it has been favorable. Users report that they are getting better mileage and smoother engine performance. In Brazil, alcohol from sugar cane has already replaced about 20 percent of that nation's gasoline use.

One of the major advantages of using solar alcohol as a fuel is that it can be produced locally with whatever is at hand. In the Midwest, spoiled grain and corn husks are being used. Elsewhere, wood chips and other waste products of lumber and paper mills can be employed. In Hawaii, as in Brazil, sugar cane is used.

Alcohol is the result of simple fermentation by anaerobic microbes. Nature has been doing it effectively for billions of years, but the process may soon become even more efficient. The field of genetic engineering is still in its infancy, but many scientists foresee the widespread use of specially tailored anaerobic microbes in commercial alcohol production. There are still many technical and legal hurdles to overcome (for example, who "owns" such bacteria once they have been created?), but the future seems promising.

Nature is also adept at producing methane from organic matter. Two prime sources for conversion to methane are manure and municipal waste. Methane from manure is especially attractive because, depending on the process used, the residue that remains after the methane has been extracted is high in protein content and can be re-fed to the stock. The Peoples Gas Company of Chicago has already begun purchasing methane produced in Oklahoma stockyards. On the other hand, the city of Seattle explored the possibility of using municipal waste to produce commercial methane, but finally rejected the idea after the projected costs rose too high.

Economics aside, there are also environmental and moral considerations standing in the way of large-scale biomass conversion. "One of the problems in the whole area of biomass energy," SERI's Don Jantzen told me, "is to make sure that we do that process in a way that is really renewable over the long time frame. It's possible, of course, to ruin soil, and there are examples around the world of civilizations that have a good agricultural system and have ruined it. . . . What we want to do is make sure we set our systems up so we don't repeat that. One of the important things is to be able to recycle the nutrients back into the soil so they aren't lost." The whole point of turning to solar energy is that it is a renewable source; biomass conversion must be handled in a way that doesn't defeat its very purpose.

It may also become difficult to justify using food for fuel. If half the world is starving, using protein sources to run automobiles may be a morally unacceptable solution to our energy problems. The prospect of vast "energy plantations" is appealing, but there may be better ways to go. It may make more sense to harvest kelp from the oceans on a commercial basis; kelp does have some nutritional uses, but it is hardly a dietary staple.

Presently, the United States derives about 2 percent of its energy from biomass. The Harvard Business School study projects an increase to about 6 percent by the turn of the century. Biomass is a field in which technological and scientific breakthroughs are possible but not probable, so the 6 percent figure seems reasonable. If the oil wells dry up sooner than expected, our dependence on solar liquid fuels could increase dramatically. In any event, biomass conversion should play a progressively more important role, and eventually it will have to replace fossil fuels completely.

SOLAR-GENERATED ELECTRICITY

The third broad field of solar energy is the conversion of sunlight to electricity. In many ways this is the most controversial area of solar research because there are radically different methods of achieving the same result. Each method carries with it a host of political and economic implications, and the choices we make now will have a profound effect on life in the twenty-first century. At stake is nothing less than the basic economic structure of our society.

There are two general ways to get electricity from sunlight. It can be done directly, through the use of photovoltaic systems, or it can be done indirectly by using solar energy to run generators. The technology involved runs the gamut from Stone Age to Space Age, and the hardware ranges in size from smaller than your fingernail to bigger than your state.

Photovoltaics

Photovoltaic cells are the product of a rapidly evolving technology. Unlike most solar technologies, this is an area in which dramatic breakthroughs are not only possible, but likely. Photovoltaics are steadily becoming more efficient and less expensive. A decade ago, photovoltaics were practical only where other forms of power could not be used effectively, and where cost was not a prime concern, as in spacecraft. Today, we seem to be on the verge of widespread use of photovoltaics for thousands of different applications.

The typical photovoltaic cell is made of silicon, selectively impregnated with impurities. When solar photons strike the cell, electrons from the impurity are set in motion toward a negative electrode, while a positive charge accumulates at the opposite electrode. An electrical current then flows between the two. The efficiency of the cell is determined, in part, by the purity of the silicon. Defects in the structure of the crystal or the presence of anomalous impurities can cause the free electrons to recombine with the crystal lattice before they reach the electrodes.

In the early 1950s, the maximum efficiency achieved in converting sunlight to electricity was only about 1 percent. That figure has climbed rapidly, mainly because of the skyrocketing growth of solid state technology. Unfortunately, photovoltaic technology and semiconductor technology now seem to be headed in opposite directions. Computer chips have been getting smaller, but solar cells must get larger to be practical in everyday applications.

Today, commercially available photovoltaic cells have an efficiency of about 14 percent. "That efficiency will not double," according to Ralph Kerns of SERI's Photovoltaics Branch, "but it is possible that we could go up by another two, maybe to five percent, depending on how the technology goes." In laboratories, a few cells have been produced with efficiencies of 19 to 20 percent, and that is getting close to the theoretical limit. Depending on the details of how a cell is built, the maximum possible efficiency is about 22 to 24 percent. "But twenty percent," says Kerns, "practically speaking, is probably an upper limit for commercial cells in single-crystal silicon technology."

There are other methods by which photovoltaic cells can be made, but there are trade-off problems of cost versus efficiency.

This array of photovoltaic cells is located on the Papago indian reservation in southwestern Arizona, and is providing the residents with electric power for potable-water pumping, lights in the homes and community buildings, family refrigerators, and a communal washing machine and sewing machine. Prior to its installation in late 1978, the 15 families in the village had no access to electric utility power.

U.S. Department of Energy. Photo by NASA

Single-crystal silicon cells are efficient but expensive. One of the major difficulties is that after the crystals have been grown, they must be cut into extremely thin wafers. Because of the strength of the silicon, the saw blade must be thicker than the wafer; thus, more than half of the material is wasted. Better techniques are being developed, including some derived from the Italian marble-cutting industry, but the problems inherent in single-crystal silicon have led some scientists to explore alternative silicon structures.

One new area of research is known as amorphous silicon. In this case, silane gas (SiH_4) is, in effect, electroplated in superthin layers on the surface of the cell. This method eliminates the rigid crystal structure, with a consequent drop in efficiency. The maximum efficiency achieved with amorphous silicon has been about 6 percent. That can probably be improved upon, and the lower production costs make amorphous silicon a viable competitor with traditional single-crystal cells.

Another approach to the problem lies in the use of different materials. Silicon in its raw form (sand) is cheap and abundant, but other materials may yield comparable efficiencies. "We have really just touched the surface of the available chemistry," says Kerns. Among the materials being considered for solar cells are gallium arsenide, cadmium sulfide, and gallium phosphide. The costs of these alternate materials poses a problem, and there is also some question about their environmental effects. "Some of the chemicals that you use in the processing of a gallium arsenide solar cell," Kerns points out, "are extremely toxic and volatile. So you do have to be careful with them, and that's one thing you avoid, at least to some extent, with the silicon technology."

The controlling factor in the use of photovoltaics is the cost. The price of installing a photovoltaic system capable of providing a home with 3 to 5 kilowatts of electricity is presently around $50,000. That price is guaranteed to come down, but to be practical, it must be reduced by a factor of 10 to 20.

Photovoltaics produce electricity at a price of about $10 to $13 per peak watt—that is, at maximum noontime efficiency. To be competitive with traditional forms of electricity production, the price will have to be brought down to a range of 20 to 50 cents per peak watt. Using amorphous silicon technology, a rate of about $2 per peak watt is possible. Some scientists predict that within a few years, the amorphous silicon cells will be able to generate electricity at a price even lower than that of conventional generators.

Dramatic cost reductions in photovoltaic technology are not merely the pious hopes of solar enthusiasts. The rate of progress has actually been much faster than anyone predicted. The Stanford Research Institute (SRI) had been engaged in a joint venture with the Jet Propulsion Laboratory (JPL) to produce a new two-step method of getting purified silicon. The goal was to reduce the current price of $60 per kilogram to $10 per kilogram by 1986. But working independently, SRI scientists announced in 1979 that they had come up with a one-step method that reduced the cost to just $5 per kilogram.

In the two-step process, powdered sodium fluorosilicate is heated to produce gaseous silicon tetrafluoride. In a second vessel, the gas is combined with sodium and heated to produce pure silicon and sodium fluoride. The one-step process uses the heat created by the reaction of sodium and sodium fluorosilicate to drive the process to completion in a single vessel. As a bonus, the by-product sodium fluoride can be sold to makers of aluminum.

This breakthrough, years ahead of schedule, bodes well for the future of photovoltaic technology. As the cost of solar electricity falls, the price of conventional electricity has been rising. Photovoltaics are already cheaper for some applications. SERI's stated goal is to reduce the cost to 30 cents per peak watt by 1990; no one will be terribly surprised if it happens sooner than that.

Since the purity of the silicon is important in determining the efficiency of solar cells, considerable attention has been paid to finding ways of producing exceptionally pure silicon. The SRI process has been successful in that regard, but some experts think that even better silicon could be produced in orbit. Working with vacuum and zero gravity, silicon of unearthly purity could be manufactured in space on board the Space Shuttle. The drawback is that the cost of transporting the raw materials up and the finished product down would add too much to the total price of solar cells. One possible solution to the problem is to leave the finished product in orbit.

Solar Power Satellites

The concept of orbiting solar power satellites is nothing less than mind-boggling. The Sun shines all day long in space and there are no clouds to get in the way. It now seems technically possible to build a series of immense solar collectors in space and beam that endless sunshine down to Earth in the form of microwaves. The fact that such a system probably *can* be built does not necessarily mean that it *should* be built. The concept is extremely controversial, and the decision to build solar power satellites cannot be taken lightly. It is, in the apt phrase of science writer J. Kelly Beatty, "the trillion dollar question."

The father of the solar power satellite (SPS) concept is Dr. Peter Glaser, of the Arthur D. Little, Inc., "think tank" in Cambridge, Mass. In 1968, Glaser suggested that in gravityless space, it should be possible to construct solar collectors several miles across. The energy could then be converted to microwaves and transmitted to huge receiving antennas ("rectennas") on the ground. The scale of Glaser's brainchild was daunting; most experts refused to take it seriously.

At about the same time Glaser was conceiving his solar satellites, Dr. Gerard O'Neill of Princeton was dreaming up a project for his freshman physics students. O'Neill asked his students a bizarre question: Is the surface of a planet the best place to live? The answer that emerged from the initial study was: No, it isn't. For a wide variety of scientific, economic, political, social, and ecological reasons, it seemed that there were real advantages in living in space. O'Neill and those who agreed with him began to dream of gargantuan orbiting space colonies, in which hundreds of thousands of people would lead an almost idyllic life.

Inevitably, the followers of O'Neill and the followers of Glaser found one another. The space colonists needed a good reason for building their sky kingdoms—that is, a reason that would be palatable to those who would have to pay the bills. The concept of a SPS system was ideal: the colonies would be needed to support the construction of the SPS system to provide limitless energy for the planet-bound populace.

There is a lot of "space cadet" psychology involved in the space colonies movement, and it detracts from the underlying seriousness of the concept. There are sound, practical reasons for investigating the feasibility of an SPS system, but the space cadets somehow make the entire venture sound like a sequel to *Star Wars*. On their list of priorities, the presumed joys of sex in zero gravity are usually mentioned at every opportunity. That certainly has its appeal, but it is not a very good reason for spending a trillion dollars.

By the early 1970s, NASA began to pay attention to Glaser's proposals. Stripped of the trappings of the trekkies, the idea seemed to make some sense, after all. NASA and DOE have now spent over $20 million studying SPS. So far, they have found no "showstoppers"—there is nothing inherently impossible in the SPS concept.

As the plan is currently envisioned, an array of sixty satellites would be constructed in low orbit, then shifted to a higher geo-synchronous orbit at an altitude of about 22,000 miles. Stationed permanently above the United States, each satellite would have a surface area of about 21 square miles and the potential to produce 5 billion watts (1 billion watts = 1 gigawatt) of electricity. The transmitters would beam a tightly focused microwave signal to the rectennas on the ground. Each rectenna would cover some 30 square miles of land, and be surrounded by a 1.5-mile-wide buffer zone to prevent the microwaves from doing environmental damage. The fleet of sixty satellites would produce a total of 300 gigawatts of electricity, which is roughly equal to the entire United States demand at the present time.

The advantages of such a system are obvious. SPS would provide essentially unlimited power, free of pollution, on a nearly continuous basis. (The satellites would be "eclipsed" twice each year for a period of about 72 minutes, but this is not regarded as a major problem.) The SPS would be the ultimate Big Fix.

However, the disadvantages are equally obvious. Militarily, no nation could afford to have its entire power supply floating in the sky like a space age shooting gallery. Environmentally, the satellites would not be quite as "clean" as envisioned. The staggering number of rocket launches required to get all the hardware into orbit could have a disastrous effect on the atmosphere in general and the ozone layer in particular. The effects of microwaves on organisms are not well known, but it might not be healthy to live near one of the rectennas. Land use is also a problem—the sixty-satellite scenario would require nearly 2,000 square miles of receivers.

None of these problems is unsolvable. In a rare display of wisdom, NASA and DOE are studying these issues intensively *now*, before it is too late to do anything about them. By comparison, twenty-five years after the first nuclear power plant went into operation, the utilities still have not figured out what to do with their nuclear waste.

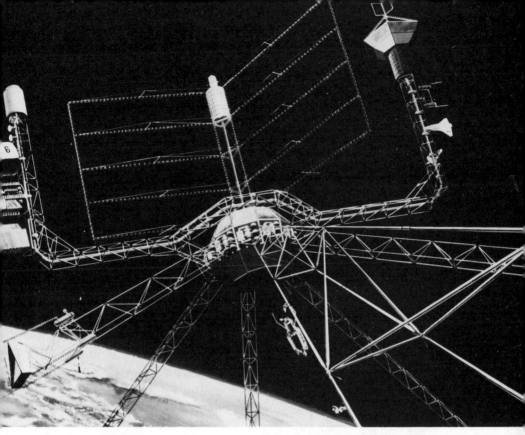

This artist's conception shows one possible configuration for a solar power satellite. Shown here is the engine section, necessary to lift the finished satellite from low orbit to its permanent station in geosynchronous orbit. The spidery legs, designed to take advantage of zero gravity, would support huge arrays of solar collectors. The scale can be seen by the size of the Space Shuttle, docked at upper right.

Boeing Aerospace Company

The SPS enthusiasts are hopeful that they can identify and neutralize the problem areas long before the first satellite is constructed. NASA and DOE scientists are leaving few stones unturned; currently, they are even studying the effects of microwaves on the birds and the bees. SPS advocates claim that directly under the rectennas there would be no microwave leakage. They picture herds of contented cows placidly grazing beneath the rectenna umbrella. But migrating birds and insects would fly right through the microwave beam, with possibly harmful effects. Ironically, new research indicates that the birds might actually *like* the microwaves—the beam apparently has a warming effect on them. That could lead to a whole new set of problems, in case the birds like the microwaves so much that they decide to build nests on the rectennas.

The problem of too many rocket launches could be circumvented by setting up mining operations on the Moon. Most of the material needed for the satellites (and colonies) is present on the Moon, and the low lunar gravity would make it relatively easy to deliver the raw materials to the assembly point in space. The space colony buffs see the establishment of a Moon mining business as an essential step in the long-range future of human civilization in space. They are not unmindful of historical analogies. After the Lewises and Clarks (or Armstrongs and Aldrins) have done their jobs, the next step is to initiate commerce in the wilderness. After the mines are in operation, the miners will inevitably be followed by wives, sweethearts, preachers, and schoolmarms. Manifest destiny takes over.

It may, indeed, be our destiny to populate the cosmos, but that cannot be considered a compelling reason for constructing the SPS. Columbus did not sail because he wanted a city in Ohio to be named after him; he had more immediate, practical motivations. The SPS concept has been around for more than a decade now, and it still seems to be a realistic possibility. It may make sense to make the commitment to a small-scale pilot project to test reality versus theory. But even a small SPS would cost billions of dollars. The entire system would cost more than $1 trillion: that's the price of 8 Vietnam Wars, 30 Apollo Projects, 200 Manhattan Projects, or 500 Panama Canals. That kind of commitment must be made on the basis of hard facts, and not simply the desire to go where no man has gone before.

Power Towers, Wind, and Water

Photovoltaic cells and solar power satellites represent the extreme ends of the solar energy spectrum. Somewhere between the two are other examples of high technology, capital intensive solar power systems. The DOE has devoted much attention and money to studies of solar power towers. The power towers consist of huge mirror arrays, focusing reflected sunlight on a receiver mounted on a tower in the middle of the mirror field. There, the solar heat produces steam which drives a turbine generator.

A functioning power tower already exists in France. It was constructed for demonstration purposes, but it works so well that the French are now using it as a source of commercial electricity. In America, a 5-megawatt test facility has been constructed in Albuquerque, N. Mex., and the Barstow Power Tower in the

California desert is expected to contribute 10 megawatts to the power grid by 1981.

Power towers seem to have a future in sunny regions such as the Southwest, but weather and land use problems will probably restrict their use elsewhere. Costs are also a prime concern. The Barstow Power Tower was built primarily with DOE money. Without such government support, it would be a very uneconomical way of producing power. The current cost per peak watt from a power tower is between $7 and $10. Even optimists doubt that the price will ever get much lower than $2 per peak watt; some photovoltaic systems are already doing better than that.

Wind and water, mankind's oldest energy sources, are also being explored as possible large-scale power producers. One concept, Ocean Thermal Energy Conversion (OTEC) would take advantage of the temperature differences in ocean currents. Warm surface waters would be used to evaporate ammonia, which would drive a generator; cold water from the deep would then be used to condense the ammonia for recycling. There are numerous variations on this scheme, most of which involve pumping enormous quantities of cold water up to the surface. The OTEC concept works best in areas where there are substantial differences in water temperature, as in the Gulf Stream.

Again, there are obvious drawbacks. Large-scale OTEC operations would require the construction of immense sea-going platforms and costly submarine cables to bring the power back to the mainland. After the initial expenses, power generation would be relatively cheap, and there is no danger of running out of ocean water. But ocean currents have been known to shift, and the environmental effects of bringing so much cold water to the surface are unknown.

Using the tides and waves to generate power is nothing new; this is already a major source of electricity in Canada. However, in most places, wave action is not intense enough to make the system practical as a primary power source. Similarly, hydroelectric power generation is ideal in some areas and destructive in others. Hydroelectric power production is not generally thought of as a solar system, but in order to get energy from falling water, the water first has to be transported uphill; solar radiation accomplishes that. Using the water as it falls back downhill is an effective and sensible way of generating electricity, but only if we keep a sense of perspective. The U.S. Army Corps of Engineers' "Dam

Electrical generating platforms that float beneath the ocean surface are being studied by the U.S. Department of Energy as one method of tapping the sun's energy stored in the sea. The Ocean Thermal Energy Conversion system shown in this artist's conception utilizes the thermal gradient between warm surface-water temperatures in the tropics and the colder deep-water temperatures to vaporize a fluid like ammonia or propane. The vapor turns a turbine, which in turn drives a generator to produce electricity. This particular unit is 250 feet in diameter, 1,600 feet long, and weighs about 300,000 tons, and would send 160 million watts of power ashore to distribution networks.

Lockheed Missiles & Space Co., Inc. U.S. Department of Energy

it all" philosophy has generated environmental disasters as well as electricity. Egypt's Aswan High Dam is another example of slow environmental suicide for the sake of electricity. Most of the power generated by the dam is used to run fertilizer plants, which are now necessary because the Nile can no longer replenish the soil with silt. Hydroelectric plants can be an important source of power, but if we expect them to provide much more than the present 14 percent of our national electricity consumption, the final price tag may be much higher than we expect.

The wind has been an underexploited energy source for most of this century. Cheap oil made windmill generators seem old-fashioned and superfluous, but the price of oil has gone up and the wind is still free. The DOE is now exploring the possibility of constructing huge space-age windmills to produce electricity in

On July 11, 1979 the world's largest wind turbine generator, with blades as long as the wingspan of a Boeing 707, was dedicated near Boone, North Carolina. At winds of 11 miles per hour the giant turbine begins to generate electricity, and at winds of 25 miles per hour or more, it can produce 2,000 kilowatts, approximately the amount of electricity used by 500 average homes.

Photo by Dick Pebody. U.S. Department of Energy

commercial quantities. Several demonstration projects already
exist, and it seems likely that more will be built. The modern
windmills consist of high towers supporting gigantic propeller
blades, some as long as 300 feet. The DOE hopes to get as much
as 3 megawatts of electricity from these behemoths. Smaller
windmills (DOE prefers to call them "wind machines") may be
more practical; "farms" of such generators could be an ideal source

This is an experimental model of a vertical-axis wind turbine, which
has been mounted atop a building at DOE's Sandia Laboratories in
Albuquerque, New Mexico. The 15-foot blades drive into the wind,
producing about three horsepower in a 20-mile-per-hour wind.

U.S. Department of Energy

of power for islands and small, remote towns. There are still some problems to be overcome, however. The modern windmills are less reliable and require more maintenance than those used by our great-grandparents. Another problem not encountered by our forebears is the fact that the metal blades of the giant wind machines tend to interfere with television reception. These difficulties are minor in comparison with those faced by other solar technologies, and it seems certain that wind power will play a much greater role in the future.

Solar research is also focusing on some less obvious ways of using energy from the Sun. It may prove possible to use solar energy to split water molecules chemically and extract the hydrogen as a fuel. Hydrogen gas seems to have good potential as a fuel of the future, but it is difficult to handle and still expensive. It also has severe public relations problems; people still remember the *Hindenburg*.

Scientists are also investigating the possibility of doing commercially what plants do naturally. "Synthetic photosynthesis" may be possible on a large scale, but natural photosynthesis is not very efficient. Getting commercial power from plants would probably require spreading thin, genetically engineered super-photosynthesizers over the surface of shallow ponds. A considerable amount of land would be needed for such "electricity farms." It is far too early to write off this concept as impractical, but for the foreseeable future the plant kingdom is not likely to become a major exporter of electricity.

It seems clear that between now and the end of the twentieth century, solar energy will provide an ever-increasing percentage of power. What forms it will take depends more on political and economic factors than on technology. The publicly stated goal of a 20 percent solar economy by the year 2000 is not unrealistic. Public opinion polls have consistently shown that most Americans favor a transition to solar energy. The unanswered question is whether the government will lead or impede that movement.

Eleven: THE POLITICS OF THE SUN

In the summer of 1979 I visited the Solar Energy Research Institute to do research for this book. My first interview was with Jerome Williams of SERI's Public Information Office. Before I could ask him any questions, he asked me, "Is this going to be a political book?" His question surprised me, but my own answer surprised me even more. "Yes," I said, "it seems inescapable."

My answer surprised me because until that moment I had been thinking about the Sun almost exclusively in scientific terms. Yes, I would deal with solar energy, and of course, solar energy was fraught with political implications, but as a science writer I supposed that I could justifiably skirt those issues. Traditionally, science writers talk about theories and hardware and such, and leave the political punditry to the syndicated columnists. But in the wake of the accident at Three Mile Island, the politicos were suddenly writing about science, because they had to. The issues could no longer be separated. Williams's question made me realize that the same forces influencing the political writers also affected science writers. And so, a book which began in the dark reaches of the solar nebula now approaches its conclusion in the smoke-filled rooms of contemporary politics.

The basic question is: How will our society produce the energy it needs to survive? Solar energy has been proposed as one possible answer. The alternatives seem to be limited, in the end, to nuclear energy and coal. Each alternative implies a different kind of society; we are not simply choosing a fuel, we are choosing a life style.

If we do choose a solar future, many of the issues mentioned in this book will become much more than mere scientific curiosities. If we are staking our future well-being on the Sun, it behooves us to find out as much about it as we can. The better we understand the Sun, the better we will be able to use it.

If we decide, instead, to rely on energy from coal or nuclear reactions, other scientific issues will enter the political arena. If we increase our reliance on coal, then it is inevitable that the atmospheric CO_2 content will also rise. The consequences of global heating are impossible to predict, but it is certain that there *will be* consequences; somehow, they must be factored into the decision-making process, and not simply as a footnote to someone's "environmental impact statement."

Nuclear energy also carries with it a basketful of consequences. After the near-disaster at Harrisburg, the general public has become more aware of the problems associated with nuclear energy, but the name-calling and finger-pointing that ensued did little to clear up the confusion. Nuclear advocates rightly emphasized that Three Mile Island, where no one was killed, was the only "major" accident in over 200 reactor-years of operation. But if that is a reliable gauge of the hazards, then in a nuclear-powered future, society would have to endure another Three Mile Island every year or two. The ready availability of fissionable material would make nuclear terrorism virtually inevitable. Nuclear waste would multiply, and we haven't even decided what to do with the waste we already have.

Solar energy promises a very different sort of future. "Solar America" would be a nation secure in its energy independence. There would be no need to send in the Marines to protect our fuel supplies. Our national wealth would not be squandered on overseas oil. The air and water would be cleaner in Solar America, and our dependence on energy from the biomass would make the preservation of the environment a high priority concern. The decentralization of energy technology would work to make society more democratic.

It should come as no surprise that the author of a book about the Sun favors solar energy over that produced by coal and nuclear fission. Lately, in fact, it seems that everyone is jumping on the solar bandwagon. Congress has been appropriating more money for solar research, politicians speak of it as an "apple pie and motherhood" issue, and even oil companies and other giant corporations are speaking guardedly about the promise of solar energy.

But everyone has his own idea of what solar energy means and how it should be used. As we enter the 1980s, there is still no coherent national energy policy, nor has there been a substantive commitment to the large-scale development of solar resources. As one SERI staffer told me, "We don't really have a policy. Nobody at SERI or on Carter's staff has really sat down and talked about how we get from here to there."

SOLAR VERSUS NUCLEAR

Part of the problem lies in the old saying that everybody wants to go to heaven, but nobody wants to die. Coal and nuclear energy, for all their drawbacks, at least offer the hope of continuing "business as usual." A complete conversion to solar energy, on the other hand, would necessitate some fundamental changes in the economic organization of our society. That prospect is not equally pleasing to all. If we are playing Monopoly (and we are), the player who owns the lucrative blue and green properties (Boardwalk and environs) is not likely to favor any restructuring of the board that features a shortcut from "Free Parking" to "Go." And solar energy could be that shortcut.

Another aspect of the problem is the public's perception of the problem's reality. Polls consistently indicate that the majority of the American people do not believe that there is an energy crisis. The rising price of oil and the gas lines are perceived as the result of machinations by oil companies and OPEC ministers. In the short run, that may well be true. But the suspicion that the current crises have been manufactured has led many people to doubt the reality of the long-term oil shortage.

In an atmosphere of uncertainty and doubt, people tend to decide by not deciding. A relevant historical analogy is close at hand. In July 1961, President John F. Kennedy alarmed the nation in a speech on Berlin by urging every American to take steps to protect his family in the event of nuclear war. In the context of the times, it seemed that Kennedy was telling people to start digging fallout shelters. In reality, Kennedy was simply trying to get public support for a tougher line with the Soviets, and the fallout shelter remark was an ill-considered attempt to get the people to take it all personally. But Kennedy had unwittingly stirred up a hornet's nest. Almost overnight, the home fallout shelter business blossomed. While national magazines ran spreads on the latest in home survival technology, theologians debated the morality of shooting your neighbor at the shelter door, and fast-buck artists thrived on scared and gullible consumers. Confusion was rampant, and when people turned to the government for guidance, they found none. Before Kennedy's speech, there had been no organized national shelter policy. Spurred by the outcry, the Administration attempted to come up with a policy in a hurry but failed completely. Finally, several months later, Kennedy almost casually mentioned that he had never meant that people should go out and dig holes in their backyards—what he had in mind was more of a community shelter program. The whole idea of fallout shelters was allowed to slip quietly into oblivion. In the meantime, with no leadership from Washington, the public reacted to the "crisis" by doing little or nothing. Very few home fallout shelters were actually built; everybody talked about it but almost no one actually did anything.

We now find ourselves in a similar situation. Our political leaders have sounded the alarm but have offered little in the way of a concrete program. Home solar systems seem to represent a personal commitment which the government is unwilling or unable to match with a national commitment. At various times the

Carter administration has spoken glowingly of the future of nuclear energy, synthetic fuels from coal, and solar energy. This "all things to all people" approach only compounds the confusion and uncertainty. People seem to be waiting for someone in authority to bite the bullet and spell out the steps that must be taken. To date, that has not happened.

The most confusing aspect of our energy situation is that it has become a "battle of the experts." During the tense days of the Three Mile Island crisis, the public watched in dazed fascination as a parade of pro- and anti-nuclear experts attempted to explain what was happening. If the experts disagreed, how was the average man to determine the truth? Similarly, pro- and anti-solar experts publicly disagree on the potential of the Sun. Some hail solar energy as the only rational alternative, while others maintain that it is an inherently limited source of energy and can never play more than a minor role in our energy future. Who is right?

Scientists, traditionally, are reluctant to get involved in political disputes. That tradition is changing, but there is still a strong feeling that a scientist's job is to pursue the truth, independent of public opinion and political pressure. Some scientists believe that they could lose that independence by taking a position on sensitive issues.

The counter-argument is that scientific independence is an illusion and always has been. Science is an expensive pursuit, and the money has to come from somewhere. Mostly, it comes from big corporations and the government. In recent years, we have seen a depressing number of cases in which scientists have tailored their findings to suit the needs of drug companies, tobacco companies, oil companies, and nuclear utilities. The tradition against scientific advocacy seems to apply only to those scientists who oppose the conventional wisdom.

In the solar versus nuclear debate, the solar advocates are generally pictured as radical hippie mavericks; the pro-nuclear scientists, on the other hand, are accused of being bought and paid for by the establishment. There is some truth in both charges—enough to make other scientists reluctant to get into the fray. But the pressure for commitment is mounting. One SERI staff member told me, "As an institution, we are not anti-nuclear. But I don't know whether we can continue to take that position."

Increasingly, scientists and institutions *are* taking an anti-nuclear position. Three Mile Island seems to have done for the nuclear industry what the *Hindenburg* did for the zeppelin industry. Nuclear power has not yet been counted out, but it is clearly in trouble. The nuclear advocates have yet to demonstrate that nuclear reactors can ever be made entirely safe; however well designed they may be, the reactors will still be operated by fallible human beings. The safety question aside, the economics of nuclear power are becoming less appealing. Building a new reactor requires a commitment of ten or more years and a capital investment of more than $2 billion. Moreover, we are facing a possible uranium shortage which could become as serious as the oil shortage.

FUSION POWER

As fission becomes less attractive, nuclear advocates are placing more emphasis on the promise of fusion power. The controlled fusion of hydrogen—the process occurring in the Sun itself—is being advertised as yet another Big Fix. There are reasons for thinking this may be true. A fusion reactor would theoretically be safe and clean, producing no unmanageable waste or meltdown alarms. The fusion fuel—water—is cheap and inexhaustible.

But controlled fusion is still only a dream. Fusion requires extremely high temperature and pressure, which have not yet been achieved in any laboratory in the world. The most optimistic projections are that small-scale fusion might be achieved by the mid-1980s. A small demonstration reactor could be constructed in the 1990s, and commercial fusion might become a reality by the second decade of the twenty-first century.

Even nuclear advocates are now tending toward a position of: "You may not like fission, but you'll love fusion." Fission is portrayed as a somewhat unpleasant but necessary part of the transition to a fusion-powered future. But the large-scale use of fusion is likely to be incredibly expensive, and there have been some indications that fusion might not turn out to be quite as safe and clean as its proponents claim. The primary drawback of fusion, however, is that it is still on the drawing boards. As Barry Commoner has pointed out, "To be capable of supporting a future, renewable system, an energy source must first meet a rather rudimentary test: it should exist. . . . To embark, today, on a transition to a nonexisting but hoped-for renewable energy source would be like building a bridge across a chasm without first locating the other side." Whatever its potential for the future may be, fusion cannot be numbered among today's energy options.

That brings us back to fission, coal, and the Sun. Of the three, only solar energy is a renewable, safe, and clean energy source. Surely, the choice is obvious.

BIG SOLAR VERSUS SMALL SOLAR

But if the choice is obvious, the manner of its implementation is anything but clear. As elaborated in Chapter 10, the use of the Sun involves a boggling variety of technologies. "Going solar" could mean almost anything. The decision to convert to solar energy is only half of the story; the way in which we do it is equally important.

On the office wall of one of the SERI scientists hangs a poster which speaks volumes: "Solar energy is harmful to oil companies and other living conglomerates." It has long been axiomatic among solar advocates that: "you can't put a meter on the Sun." That may be literally true, but it has not prevented the corporate powers from trying. They may even succeed.

There are two basic scenarios for a solar society. In one version, we follow what physicist Amory Lovins refers to as "soft energy paths." Every home has a solar heating system and a photovoltaic array. Backup power is provided by wind, biomass methane, and co-generators utilizing waste heat. Transportation relies on solar liquid fuels, such as alcohol. Exxon and Consolidated Edison wither away in the heat of the Sun.

The second scenario is more familiar. The average consumer still buys his heat and electricity from the utilities and his fuel from the oil companies. The power grid is fed by energy from commercial power towers and solar satellites, instead of oil and coal. The monthly bill for it all remains devastatingly high—somebody has to pay for those satellites and towers.

Either scenario seems preferable to a nuclear- or coal-powered future. The real choice is between a society in which the Sun belongs to everyone and one in which it belongs to the same people who already own the oil, coal, and uranium.

In the current structure of our economy, virtually everyone must purchase energy from a centralized producer. In the Big Solar scenario, that situation would be essentially unchanged. In the Small Solar scenario, however, individual consumers would get the bulk of their energy directly from the Sun, with no middleman involved. Centralized producers would be relegated to no more than a backup role.

A complete conversion to a Small Solar society would break the economic stranglehold of the giant corporations. Obviously, the executives of Exxon, Shell, General Electric, *et al.*, would not be overjoyed by such a turn of events. Their actions in recent years strongly suggest that corporate leaders have evolved a two-pronged strategy for dealing with the solar threat. In the first place, solar energy is minimized as a serious power source. The second part of the strategy is to gain control of whatever solar development does take place.

A reader whose only knowledge of solar energy was provided by advertisements paid for by Exxon, Mobil, and other large corporations might well conclude that solar energy is small potatoes, indeed. A series of Exxon advertisements in 1976 offered drastically inflated cost figures for home solar users and suggested that solar energy "possibly" could become a major energy source "in the next century." The ad was such a flagrant distortion of solar realities that Senator Gary Hart and Representative Richard Ottinger wrote an indignant response to Exxon President H. C. Kauffmann. "We are left to wonder," wrote the Congressmen, "whether the answer to the final question (in the ad), 'Why is Exxon involved in solar energy?' might simply be that Exxon intends for solar energy to be kept under wraps until fossil fuel markets are exhausted." Similar advertisements have been sponsored by Shell and Mobil.

Another dubious exercise in "public education" was conducted by Honeywell, Inc., under a federal grant. Honeywell created something known as the Transportable Solar Laboratory, which was basically a "show-and-tell" presentation packed into a mobile home. The TSL traveled around the country and explained the facts of solar life to the public. In the Honeywell version, solar energy was undependable and expensive, and could only be developed by the genius of giant corporations. The TSL show was so blatantly slanted that the California State Energy Commission urged that the tour be "immediately stopped and never started again."

The purpose of such propaganda seems to be the reduction of public demand for solar energy. If the public perceives solar energy as an expensive, unreliable alternative, there will be little demand for solar development. The effect is evident in the attitudes of architects and builders. "Both [groups are] very conservative," says SERI's Bruce Baccei. "I think that one of the biggest forces for them to change will be public demand. In any area where you're trying to push for change, you have to balance development of professional expertise with demand by the public. It's really interesting. You go to a building manufacturer and say, 'Why don't you incorporate passive solar in the buildings you're producing?' And they say, 'Well, there's no demand.' And then you say, 'Okay, I'm going to go out on a big public campaign'— and they say, 'Don't do that! We don't know how to do it yet!' That's true everywhere." Thus, we wind up with statements such as this one from the National Association of Home Builders: "Appraisers aren't recognizing solar energy as a great home improvement nor are home buyers recognizing it yet. So you don't get your money back when you sell the house you've put a solar system into."

As long as public demand remains low, solar energy will be more expensive than it has to be. Without the economies of scale made possible by mass production, home solar systems will inevitably cost more than conventional heating systems. The big corporations have a vested interest in keeping things that way. If I buy a solar heating system and cut my monthly energy bill in half, that costs the utilities about $200 a year. If a million people do the same, the cost to the utilities is $200,000,000. If the energy producers can keep those million consumers convinced that solar energy is ineffective, their anti-solar ad campaign will net them a cool $1 billion in five years.

The skeptic, at this point, is entitled to ask, "So what? Being anti-solar makes money for the corporations, but couldn't they make a lot more by *creating* a public demand and then manufacturing solar systems?" The answer is yes, they could, and yes, they probably will. But it is a question of timing. With oil prices now in the range of $30 to $40 per barrel, why kill the source of the golden eggs? If solar heating has the potential to replace 3 million barrels of oil per day, it makes sense (for the oil companies) to delay the realization of that potential as long as they reasonably can. If those 3 million barrels were magically eliminated today, the gross cost to oil producers for a single year would be about $36.5 billion. Even the Arabs could not simply blink at such a loss.

Sooner or later, the oil wells will run dry. Long before that happens, the price of oil will probably reach such a high level that the advantages of solar energy will be impossible to ignore, even without mass production. When that day arrives, the major corporations will be ready.

The national solar energy policy, such as it is, has been heavily influenced by the major corporations. A laundry list of the firms involved in developing government solar policy would include names such as IBM, RCA, AT&T, TRW, Alcoa, General Electric, Dow, Dupont, Mobil, Shell, Exxon, Gulf, Bechtel, Boeing, Martin Marietta, General Motors, Chrysler, U.S. Steel, and a handful of major banks. Whatever solar policy is finally chosen, it is unlikely to be harmful to the interests of those who create the policy.

Again, we can ask, "So what?" It can be—and has been—argued that what's good for General Motors really is good for the country. The giant corporations got to be giants by doing whatever it is that they do, profitably and with reasonable efficiency and competence. Despite an occasional DC-10 or Pinto, the industrial complex generally produces good hardware. If I'm in the market for a solar heating system, do I want to buy it from General Electric, or from Ed's Solar Plumbing Shop?

It's at least possible that I may want to buy it from Ed—if his product is better. The American tradition of bicycle-shop genius is not dead, and solar energy has been a fertile field for latter-day Orville Wrights. Small-scale operations in virtually every area are consistently more innovative than big budget research and development laboratories. They have to be, or they don't survive.

In developing solar energy technology, the DOE has generally ignored the small companies in favor of the *"Fortune* 500" giants. Honeywell gets $1.1 million for its transportable travesty, but a tiny company in Texas can't get $100,000 for a demonstration of a promising new method of using solar energy to generate electricity. A man in Minnesota who claims to have developed a practicable black solar fluid is given a runaround, while Batelle Research Institute is given $125,000 to develop the same fluid. Similar horror stories abound. Small Solar advocates argue that they can't get federal money to develop their ideas and that, if their ideas are already developed, the DOE ignores them or subsidizes competition by the big companies.

The DOE insists that this is not true. In each project, a certain percentage of the funding is earmarked for small businesses. When a Request for Proposals (RFP) is sent out, the door is open to all. Further, there is an inherent risk in dealing with small companies. As Chuck Kutscher of SERI's Thermal Conversion Branch put it, "When you're dealing with businesses that may or may not be around for the whole period of time of the contract, things get a little bit scary."

There is, however, a built-in bureaucratic bias that works against the small companies. Proposals for government funding often require thousands of man-hours to put together, and small operations simply don't have the time, manpower, or expertise to compete with the heavyweights. Critics also complain that the small companies are frequently not even notified that an RFP has been sent out.

The overall direction of national solar energy policy is such that most of the funding goes to large-scale capital-intensive Big Solar projects which small companies simply are not equipped to handle. A case in point is the Barstow Power Tower, which SERI's new Director Denis Hayes in a "Sixty Minutes" interview labeled "a gold-plated turkey." The Power Tower has consumed more than 180 million federal dollars, most of which went to corporate giants, such as Honeywell, McDonnell Douglas, Boeing, and General Atomic. Ed's Solar Plumbing Shop can't compete.

The corporate solar policy, faithfully reflected in the government solar policy, seems to be that to the extent the nation goes solar, it should go Big Solar. This not only excludes competition from small innovators, it also assures that the control of energy resources will remain in the hands of the big corporations. If you can't put a meter on the Sun, you *can* put a meter on a power tower, a megawatt wind machine, or a solar power satellite. Any advances in Small Solar technology represent a threat to the corporate game plan, even if those advances are produced by "members of the club."

The Jet Propulsion Laboratory, operated by California Institute of Technology, is one of the centers of photovoltaic research. Photovoltaics can be either Big Solar or Small Solar, depending

on how they are employed. A recent JPL study illustrates the Big
Solar bias.

I met with Dr. Nevin A. Bryant in his (inevitably) tiny office
at JPL, where he explained the study. There had never been a
direct comparison of actual power use versus solar power production on a large scale. To investigate this important subject, JPL
employed spacecraft and computer analysis techniques to examine
power use in a section of the San Fernando Valley, north of Los
Angeles. Initially, the researchers identified the available flat and
south-facing rooftop area within the target district. Then they calculated the potential power production, assuming that half of
the available rooftop space was covered with photovoltaic cells
with an efficiency of 10 percent. This result was compared with
the actual power use in the area throughout 1978. The result:
photovoltaic cells could provide 52 percent of the area's overall
power needs and could actively produce a surplus during the daylight hours.

The 52 percent figure is greater than the expected growth in
energy demands projected for this high-growth area over the next
25 years. There are some questions concerning the problem of
storing surplus energy and the assumed efficiency of conversion
from DC to AC for the power grid. When I spoke with Bryant,
the final report had not yet been written. "But what it's telling
us," said Bryant, pointing at sheets of raw data, "is that the potential is *enormous*."

I suggested that the result was in flat contradiction to the asser-
tions of utilities and oil companies. Their claim has consistently
been that any form of large-scale solar energy use would have to be
centralized. "That's total bullshit," Bryant responded. "I totally
disagree with it."

Bryant conceded that there were "valid quibble points" in the
study, but the central thrust of it was clear: the use of photovoltaics can make some local districts *exporters* of electricity. "There's
lots of surplus at any particular time," Bryant went on. "So how
do you handle that? This creates a whole lot of problems—I prefer to think of them as challenges—for electrical engineers. You
no longer have a straightforward system of centralized power generation . . . you have a situation where there is a distributed energy
supply as well as distributed energy demand." The study is being
expanded to produce a better "model" of the day-to-day energy

flux within the system, but regardless of the precise details, the implications are obvious. "What this irrefutably says," according to Bryant, "is that the increased power production requirements in the next twenty-five to thirty-five years in existing areas could probably be met by solar photovoltaics. . . . You wouldn't have to buy a whole new set of power plants."

I asked Bryant if he had received any reaction to the study. "We've kept this at a very low profile," he replied, lowering his voice, "because these are very provocative statistics. . . . Interestingly enough, JPL has received no funding from DOE to pursue this, even though we've requested it. So that might tell you . . ." Bryant trailed off, leaving me to decide for myself what it might tell me.

What it does tell me is that DOE has very little interest in any solar system which does not require centralized power generation. That is the antithesis of the Big Solar philosophy. In the eyes of government and big business, small is *not* beautiful.

So there it stands. Slowly, grudgingly, the nation is moving toward the increased use of solar energy. It seems likely that we will also increase our use of coal and of nuclear energy—in whatever form—and that in the twenty-first century we will depend on a hodgepodge of energy sources and technologies, all of which will be controlled by the same people who control them today. That prospect is not equally distasteful to everyone. The increase in corporate control of life's essential needs is nothing new; indeed, it is probably the most important and pervasive economic and social development of the twentieth century. Centralization certainly has its benefits.

But the Sun *does* belong to everyone. You and I can use it if we choose, without asking anyone's permission. If we are truly a nation of sheep (or children, as Nixon believed), then we will meekly accept whatever is granted us by the barons of Big Solar, Big Nuclear, Big Government, Big Business, Big Everything. But if we are the independent, self-reliant, freedom-loving people we like to think we are, then solar energy offers us a unique opportunity to decide for ourselves what kind of lives we want to lead. When the Sun comes up tomorrow, it will shine equally on all of us.

Glossary

Biomass: any material in the environment produced by living organisms. Strictly speaking, this would include coal and oil, but in terms of energy production, the biomass is usually thought of as "recently arrived" solar energy.

CNO Cycle: the Carbon-Nitrogen-Oxygen fusion chain. Hydrogen nuclei combine successively with carbon-12, nitrogen-13, carbon-13, nitrogen-14, oxygen-15, and nitrogen-15 to produce a carbon-12 nucleus, a helium nucleus, and energy. The Sun gets about 2% of its energy from this process.

Fraunhofer Lines: dark lines in spectra produced by the absorption of energy at particular wavelengths by atoms or ions.

Gaia Hypothesis: a theory that states that the Earth's atmosphere behaves as if it were a huge living organism, constantly making adjustments to keep the environment suitable for life.

Greenhouse Effect: the retention of solar heat in the atmosphere of a planet due to absorption by carbon-dioxide molecules. The planet Venus seems to be an extreme example of the greenhouse effect.

Hertzsprung-Russell Diagram: a plot of absolute stellar magnitude versus spectral class. Originally devised by Ejnar Hertzsprung and Henry Norris Russell around 1913.

Maunder Minimum: an era of extremely low solar activity, notable for a nearly total absence of sunspots. It occurred between 1645 and 1715 and roughly coincided with the so-called Little Ice Age of the seventeenth century.

Milankovitch Cycles: periodic defects in the Earth's orbit, first identified by Yugoslavian scientist Milutin Milankovitch in the 1930s. The orbital defects follow cycles of 100,000, 41,000, and 23,000 years, and may be responsible for episodes of terrestrial glaciation.

OTEC: Ocean Thermal Energy Conversion. The concept involves utilizing the temperature difference between cold deep ocean water and warm surface currents to produce energy.

Proton-Proton Chain: the basic hydrogen fusion reaction producing energy in the core of the Sun. Two protons fuse to form a deuteron, which then captures a proton to become a helium-3 nucleus. Two helium-3 nuclei fuse to produce a helium-4 nucleus, two protons, and energy.

Rest Mass: the mass of a subatomic particle when not in motion. For the neutrino, the rest mass was (until recently) presumed to be zero.

SEP: Solar Electric Propulsion. Solar energy is used to eject mercury ions for spacecraft propulsion, providing low but constant thrust.

SERI: Solar Energy Research Institute. The national center for solar energy research, located in Golden, Colorado.

SNU: Solar Neutrino Unit. In the Brookhaven neutrino experiment, one SNU was defined as a neutrino capture rate of 1.0×10^{-36} captures per target atom per second. The predicted result was 5.6 SNU's; the actual experimental results have an upper limit of 1.7 $\pm$ .4 SNU's.

Solar Constant: the total amount of solar energy received at the mean Sun-Earth distance in the absence of the atmosphere, usually expressed in watts per square meter. The most probable value of the Solar Constant, S, is taken to be 1373 $\pm$ 20 watts per square meter.

Solar Cycle: generally thought of as the 11.2-year sunspot cycle, it is actually a 22-year cycle, involving two reversals of the Sun's magnetic polarity.

Solar Wind: the emission of subatomic particles from the Sun. Most of the solar wind comes from long-lived coronal holes above the poles of the Sun. The particles travel at an average speed of about 400 km per second.

SPS: Solar Power Satellites.

Zeeman Effect: the splitting or blurring of spectral lines due to the presence of a strong magnetic field. The Zeeman Effect is particularly evident in the spectra of sunspots.

A Note on Sources

Chapters 1–3

The information on stellar and solar evolution and behavior is derived from a variety of sources. Among the sources that proved most useful were: John A. Eddy, *A New Sun: The Solar Results from Skylab*, Washington, D.C., 1979, NASA SP-402; James Cornell and E. Nelson Hayes, eds., *Man and Cosmos*, New York, Norton, 1975; Eugene H. Avrett, ed., *Frontiers of Astrophysics*, Cambridge, Mass., Harvard University Press, 1977; Thornton Page and Lou Williams Page, eds., *The Evolution of Stars*, New York, Macmillan, 1968; Laurence H. Aller, *Atoms, Stars, and Nebulae*, rev. ed., Cambridge, Mass., Harvard University Press, 1971; Donald H. Menzel, *Our Sun*, rev. ed., Cambridge, Mass., Harvard University Press, 1959; George Gamow, *A Star Called the Sun*, New York, Viking, 1964; John C. Brandt, *Introduction to the Solar Wind*, San Francisco, Freeman, 1970. Also valuable were articles by A. G. W. Cameron and E. N. Parker, both in the September 1975 *Scientific American*.

Chapters 4–6

Much of the information on the problems of contemporary solar physics came from my interviews with the scientists mentioned in the text. Other sources include papers presented at the Conference on the Ancient Sun, co-sponsored by NCAR, NASA, and the National Science Foundation, at Boulder, Colorado, in October 1979; Oran R. White, ed., *The Solar Output and Its Variation*, Boulder, Colo., Colorado Associated University Press, 1977; John N. Bahcall and Raymond Davis, Jr., "Solar Neutrinos: A Scientific Puzzle," *Science*, 23 January 1976; J. A. Eddy *et al.*, "Anomalous Solar Rotation in the Early 17th Century," *Science*, 25 November 1977; John A. Eddy, "The Case of the Missing Sunspots," *Scientific American*, May 1977.

As a general source for current developments in this field (and many others), I strongly recommend *Science News*.

Chapters 7–9
On the confusing and often controversial subject of solar-terrestrial interactions, my sources included conversations with a number of scientists, mostly at NCAR and NOAA; literally dozens of publications produced by those organizations; and a variety of books and articles, especially: John R. Herman and Richard A. Goldberg, *Sun, Weather, and Climate*, Washington, D.C., 1978, NASA SP-426; Louise B. Young, *Earth's Aura*, New York, Knopf, 1977; Nigel Calder, *The Weather Machine*, New York, Viking, 1976; Norman J. Rosenberg, *Microclimate: The Biological Environment*, New York, Wiley, 1974; Isaac Asimov, *Photosynthesis*, New York, Basic Books, 1968; J. D. Bernal, *The Origin of Life*, New York, World, 1967; Readings from *Scientific American*, *Life: Origin and Evolution*, San Francisco, Freeman, 1979; Nigel Calder, "Head South with All Deliberate Speed," *Smithsonian Magazine*, January 1978; numerous papers and articles by NCAR's Stephen H. Schneider and colleagues; G. C. Reid, *et al.*, "Influence of ancient solar-proton events on the evolution of life," *Nature*, 22 January 1976; and Steven Businger, "Investigation of Possible Sun-Weather Relationships," NCAR Cooperative Thesis No. 49.

Chapters 10, 11
The information on solar energy was derived principally from my interviews at SERI and from various publications of that organization. Other sources include: Farrington Daniels, *Direct Use of the Sun's Energy*, New York, Ballantine Books, 1977; Robert Stobaugh and Daniel Yergin, eds., *Energy Future*, New York, Random House, 1979; Barry Commoner, *The Politics of Energy*, New York, Knopf, 1979; Ray Reece, *The Sun Betrayed*, Boston, South End Press, 1977; Scott Denman and Ken Bossong, "Power Politics: Big Business," *Seven Days*, June 29, 1979; a wide variety of current articles and news reports; and the manuscript of J. Kelly Beatty's article "Solar Power Satellites: The Trillion Dollar Question," *Science '80*, December 1980.

Aside from specific sources, I benefitted greatly from conversations with my fellow science writers—in particular, David Salisbury, Jonathan Eberhart, Mitch Waldrop, and Kelly Beatty. Of course, the opinions expressed in this book are my own and are not necessarily shared by any of the scientists and writers cited here.

Index